T0310154

Static Conceptual Fracture Modeling

Static Conceptual Fracture Modeling

Preparing for Simulation and Development

R.A. Nelson

Broken N Consulting
1444 New Ulm Road
Cat Spring, TX, US, 78933

Registered Office(s)
John Wiley & Sons, Inc., 111 River Street, Hoboken, NJ 07030, USA
John Wiley & Sons Ltd, The Atrium, Southern Gate, Chichester, West Sussex, PO19 8SQ, UK

Editorial Office
9600 Garsington Road, Oxford, OX4 2DQ, UK

For details of our global editorial offices, customer services, and more information about Wiley products visit us at www.wiley.com.

Wiley also publishes its books in a variety of electronic formats and by print-on-demand. Some content that appears in standard print versions of this book may not be available in other formats.

Library of Congress Cataloging-in-Publication data has been applied for:
9781119596936 [hardback]

Cover Design: Wiley
Cover Image: Photo courtesy of S. Serra

Set in 10/12pt Warnock by SPi Global, Pondicherry, India
Printed and bound in Singapore by Markono Print Media Pte Ltd

10 9 8 7 6 5 4 3 2 1

I dedicate this volume to my professional society of choice during my career, The American Association of Petroleum Geologists. I have certainly been a member of other societies, but the AAPG is the one that I have supported over 40 years and that has supported me over that period as well. I have been an author, course instructor, reviewer, Distinguished Lecturer, Committee Chair, Trustee Associate and Executive Committee member over the years and have learned how to be an active professional from the example of other AAPG members.

Contents

Foreword

The study of natural fractures can be viewed as a progression from basic qualitative observation, to detailed measurement, to analysis of the collected measurements, to use of the analyses and collected data. Early fracture studies focused on understanding the deformation of rock as recorded by the development of fractures, but since the middle of the 20th century fractures have also been recognized as important controls on fluid flow in fractured media, including hydrocarbon reservoirs. Early fracture data-collection and analyses techniques, developed for structural purposes, were not always applicable to the later purposes of understanding fluid flow. This volume, however, bridges that gap, offering insights and techniques useful in fracture analyses specific to hydrocarbon reservoirs, and provides methods for adapting those insights and the collected data to their ultimate use in modeling naturally-fractured reservoirs.

Ron Nelson has had a long and productive history in the hydrocarbon industry, having early appreciated the need to understand natural fractures and contributing significantly to solving the related problems. He studied under the pioneers of the field (John Handin, David Stearns, Mel Friedman, John Logan, Bob Berg) at the Center for Tectonophysics, Texas A&M, writing his dissertation in 1975 on fracture characteristics in strata of the Colorado Plateau (Nelson, 1975). His history of 26 years at the Amoco Oil Company made him a pioneer in his own right, and included practical problem-solving in improving recovery from specific fractured reservoirs as well as more widely-applicable research on the problems associated with fractured reservoirs. Since leaving Amoco almost two decades ago he has become a highly-respected consultant to the hydrocarbon industry and a widely-recognized fracture expert whose services are in demand around the world. Throughout his active work schedule, he has also found time to publish extensively on fractures, including writing and later revising an acclaimed volume on naturally fractured reservoirs (Nelson, 2001). In doing so he has generously and openly shared his knowledge with industry and students.

This volume is a practical, useful, and in-depth source for geologists who need to learn about fracture data-collection from the important sources presently used by the hydrocarbon industry (seismic, core, outcrop, image logs, engineering, etc.). It provides a much-needed link between fracture data collection and fractured-reservoir modeling.

John C. Lorenz, FractureStudies LLC
New Mexico, February, 2019

Symbols and Abbreviations

AFI	Average Fracture Intensity
ATV	Acoustical TeleViewer
AVO	Amplitude Vs Offset
azi.	azimuth
BB	Borehole Breakouts
BHI	Borehole Image Log
BHTV	Borehole Television
CBIL	Circumferential Borehole Imaging Log
CMI	Compact Micro Imager
CSD	Cross-Strike Discontinuities
CT	Computed Tomography
DCFM	Dynamic Conceptual Fracture Model
DEM	Digital Elevation Models
DIF	Drilling-Induced Fracture
DFM	Discrete Fracture Model
DFN	Discrete Fracture Network
DSI	Dipole Shear Sonic Imager
DST	Drill Stem Test
EUR	Estimated Ultimate Recovery
FASTCAST	Fast Circumferential Acoustic Scanning Tool
FI	Fracture Intensity
FIC	Fracture Intensity Curve
FIH	Fracture Intensity Height
FMI	Formation Micro Imager log
FMS	Formation Micro Scanner
GPS	Global Positioning System
GVR	GeoVision Resistivity
IC	Inhomogeneity Coefficient
IP	Initial Potential

KOC	Kuwait Oil Company
LWD	Log While Drilling
MFI	Mean Fracture Intensity
Mode I	Opening-Mode fracture (opening perpendicular to plane)
MS	Micro Scanner
OBMI	Oil Based Micro Imager
PI	Productivity Index
PLT	Production Logging Tool
QC	Quality control
RAB	Resistivity At the Bit
RMS	Root Mean Square
SCFM	Static Conceptual Fracture Model
SH_{max}	Maximum horizontal *in-situ* stress component
Sh_{min}	Minimum horizontal *in-situ* stress component
SSCAN	Sonic Scanner
STOIP	Stock Tank Oil in Place
Sv	Vertical *in-situ* stress component
UBI	Ultrasonic Borehole Image log
UCS	Unconfined Compressive Strength
XRMI	Extended Range Micro Imager
σ	Principal effective stress component
ε	Strain component
ϕ_f	Fracture porosity
ϕ_r	Rock or matrix porosity
k_r	Rock or matrix permeability
k_f	Fracture permeability
e	Fracture aperture
e_h	Hydraulic fracture aperture
e_m	Mechanical fracture aperture
D	Fracture spacing
G	Rigidity Modulus or Shear Modulus
E	Young's Modulus
γ	Poisson's Ratio
ρ	Density
L	Length
F	Force
g	Acceleration due to gravity
μ	Viscosity
Ls	Limestone
Ss	Sandstone

Acknowledgments

I have certainly had a long career, and over that time have benefitted much from interactions with colleagues and companies I have worked with. Indeed, the 50 companies I have either worked for, or consulted with, have given me a broad exposure to structural geology and fractured reservoir endeavors world-wide. In addition, the numerous geologists, geophysicists, petrophysicists, and engineers in those companies have taught me much about both technical issues and about how to be an effective team member in our industry. Without these experiences this current volume would not have been possible.

1

Purpose and Scope

The purpose of this manuscript is to provide a guide for the construction of a quantitative Static Conceptual Fracture Model (SCFM) from predominantly physical descriptive rock data, which along with a Dynamic Conceptual Model (DCFM) constructed from predominantly fluid and reservoir engineering data, can be used for reservoir flow simulation (concept from Trice 2000). These simulations constrain the current reservoir behavior, as well as predict how it will perform in the future.

This manuscript will discuss the various parameters that are needed to constrain the SCFM, and later, populate computer models used to generate gridded fracture models as input to simulation. The various parameters will be detailed along with techniques I have used to gather the needed data and populate the computer models. The parameters discussed are the same regardless of the simulation modeling style used or the computer programs used to house the data and make the needed reservoir calculations. In addition, I will comment on what to do and not do in technical planning when acquiring some of the needed data sets.

An important aspect of the modeling process is, in my mind, innovation. The rock and fracture data needed for the models can come from different sources and different scales of measurement. For example, constraining fracture corridor width in the subsurface can come from core, image logs and geophysical data. All three are measured at different scales and levels of precision, therefore, giving different values and accuracy. The important thing in the modeling is to make the measurement however you can with the data you have. The variation within the distribution of the measures is perhaps more important than the actual values, as values can be shifted in bulk during history matching to obtain credible results with respect to reservoir response. The guiding principle in my mind is innovation. Get the parameter distributions however you can with the data available.

Several of the topic areas described herein have previously been documented in the two editions of a previous textbook (Nelson 1985, 2001). However, this

Static Conceptual Fracture Modeling: Preparing for Simulation and Development,
First Edition. R.A. Nelson.
© 2020 John Wiley & Sons Ltd. Published 2020 by John Wiley & Sons Ltd.

**Guiding principles for constructing a usable
Static Conceptual Fracture Model (SCFM).**

1. Direct observation of the fracture system being modeled is of paramount importance (this includes core, outcrop, and borehole image logs).

2. Multiple data set sources (geological, geophysical, and engineering) must be used to constrain individual modeling parameters.

3. A well-constrained or mature SCFM is important in reducing risk in all aspects of a project's history from initial exploration through primary and secondary development.

4. The various parameters applied in creation of the SCFM require a multi-disciplinary team of workers coordinated by a study "Champion" who guides the data generation of individual team members and assists with data integration and provides a vision of the ultimate model.

Figure 1.1 A compiled list of guiding principles for constructing a usable Static Conceptual Fracture Model (SCFM).

author is not a computer modeler. Rather, I have more than 40 years' experience studying fractured reservoirs and providing real-time assistance to the experienced computer modelers during the process.

The computer modelers know very well the ins-and-outs of inputting the data and successfully getting appropriate results in the proper format. However, a team of multidisciplinary workers (geologists, geophysicists, engineers, petrophysicists, etc.) are needed to generate the basic fracture and reservoir data. This team is intimately familiar with the input data and knows its strengths and weaknesses which is especially important during history matching of simulation results. They, or their representative (a Fracture Champion), are the appropriate people to assist the computer modeling expert(s) in evaluating and selecting appropriate input of the natural fracture-related data. An excellent example of the process is given in Richard et al. (2017).

What is presented here is a procedure for completion of a well-constrained SCFM. Of course, individual reservoir studies may lack many of the data types needed for the complete model, especially early in the history of a project. In most studies, the SCFM generation occurs in steps over time. We move from early model descriptions with little hard data to later models with richer data bases to draw from, therefore, leading to a better constrained model with lower associated risk.

The procedure detailed in this book has been created using several guiding principles. These include the following in Figure 1.1.

2

What Is a Static Conceptual Fracture Model and Why Do We Build It?

Current practices in the numerical simulation of fractured reservoirs rely on the creation of both static and dynamic conceptual models to construct an integrated reservoir model to be simulated. The Static Conceptual Fracture Model (SCFM) focuses on the physical description of the fracture system present and how it varies throughout the reservoir; while the Dynamic Conceptual Fracture Model (DCFM) focuses on the reservoir fluid properties and the fluid-flow characteristics of the fractures and matrix and their variation in 3D. The first usage of the terms or model sub-types that I know of is from Robert Trice in 2000 at an SPE Research Forum on Naturally Fractured Reservoirs in Nice, France, Figure 2.1. These conceptual models are an elemental volume representation of the reservoir representing the entire reservoir with all its variability; Davy et al. (2018), Richard et al. (2017).

There is great variation in how SCFMs are created due to data limitations and scales of observation. Ideally, the model includes a quantitative description of all elements of the fracture system and their variation in 3D.

In my studies, I developed an organized series of steps I use to create an SCFM.

To complete the process for the creation of the SCFM, there are 13 major topic areas that must be constrained in the reservoir, Figure 2.2. This list highlights major study areas, each of which may contain multiple data sets, analysis techniques and computer applications. A similar list can be constructed for creation of a DCFM, but that is not included in this volume.

Highlighted in green in Figure 2.2 (first four parameters) are the standard approaches performed in a typical qualitative fractured reservoir study with focus on exploration. These standard approaches are generally anecdotal and poorly quantified and represent much of the early work in my career.

The substance of this list of topic areas is the general organization for the remainder of this manuscript. Details of the analyses of several will be described, with examples shown. There are several uses of the detailed SCFM once it is created, including exploration predictions, well placement, reserve calculations, recovery factor and reservoir flow simulation over time. The

Static Conceptual Fracture Modeling: Preparing for Simulation and Development,
First Edition. R.A. Nelson.

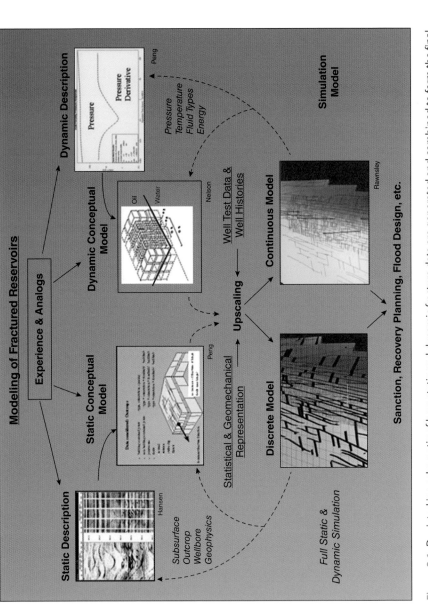

Figure 2.1 Depicted is a schematic of how static and dynamic fracture data are generated and combined to form the final simulation model; after Trice (2000), SPE Research Forum, Nice, France. This was a very early introduction to the idea of parallel Static Conceptual Fracture Model (SCFM) and Dynamic Conceptual Fracture Model (DCFM) creation that are merged into the final product for simulation.

1. Fracture Origin
2. Fracture Orientation(s) & its' distribution
3. Fracture Spacing & its' distribution
4. Fracture Morphology/Fill
5. Selection of static conceptual fracture modeling style
6. Assign fractured reservoir classification & simulation style
7. Fracture Aperture
8. *In situ* Stress Component magnitude & direction
9. Fracture Scaling
10. Predictions of properties away from control, including mechanical fracture predictions in 3D & competing mechanisms
11. Calibration to production, testing and water-cut
12. Flow Exchange (Cross Flow, Sigma Factor, Transmissibility)
13. Relation between fracture distribution and flow/ permeability

Figure 2.2 The 13 key topic areas needed to fully constrain an effective SCFM. The topic areas highlighted in green are typical for most standard fractured reservoir studies. The topic areas highlighted in blue are the additional topic areas needed for an effective quantitative fracture model. *Source:* from Nelson (2011b).

reservoir simulation application is the focus of much current work in the oil and gas industry today.

The SCFM is the database, along with a DCFM, from which gridded models for reservoir simulation are created. The data from the fracture model is input into the simulator in a specific way, depending on the modeling software (such as Petrel™, FracMan®, and other in-house company programs). Figure 2.3 depicts the general form of natural fracture parameters needed as direct input to a simulation model.

If reservoir simulation is the final product of our studies, we must keep these parameters in mind throughout the study of the natural fracture system and the creation of the SCFM.

Done correctly, data input occurs with the fracture analyst or study "fracture champion" sitting with the numerical computer modeler at the workstation, while making input decisions together. This cooperative work is necessary because the fracture analyst knows the details and strength of all the fracture databases, developed by the technical team, and the computer modeler knows how the simulator responds to that input. The gridded SCFM is then generally handed over to the reservoir simulation engineer to perform the simulation and do the first history matching of results.

The model created can be either "non-discrete," using permeability multipliers in various directions to model the effect of fractures, or "discrete" where individual fracture and structural features are placed at specific points in the model, Figure 2.4. In addition, the discrete modeling approach can use deterministic data (features placed at points in the model exactly where we observed them), or stochastic data (features placed through the model in a statistical manner based on our combined fracture system statistics).

Natural Fracture Parameters Needed for Input to Reservoir Simulators

- Dominate fracture set(s)
- Fracture set orientation
- Fracture set orientation dispersion (variability in orientation)
- Fracture set spacing
- Fracture set spacing distribution
- Fracture set length
- Fracture set height
- Fracture set aspect ratio
- Fracture set aspect ratio distribution
- Fracture set aperture
- Fracture set aperture distribution (by orientation and in map view)
- Fracture set to fracture set crossing rules
- Variation of all above spatially (maps and sections)
- Calibration of all above with well test data
- Interpretation of fracture system origin(s) for distribution prediction
- Fracture transmissibility or fracture/matrix cross flow

Figure 2.3 The natural fracture input data needed for a reservoir simulation of a dual-porosity or dual-porosity/dual-permeability reservoir. These data come from a well-constrained Static Conceptual Fracture Model (SCFM). Compiled with the help of S. Bayer in 2011, personal communication.

**Static Modeling Options For
Fractured Reservoir Simulation**

• Non-Discrete Fracture Models

Treats fractures via local matrix permeability multipliers in x, y & z and utilizes single porosity simulation.

• Discrete Models

Places individual features with independent porosity & permeability in x, y & z.

– Deterministic

Individual placement via detailed integration & mapping

– Stochastic

Statistical placement of features

Figure 2.4 Shown is a description of the types of models used to model a reservoir for simulation, based on the format of the data to be applied to the modeling. In simulating a large field, we can merge multiple regional models some of which are deterministic and other stochastic and based on the fracture database developed.

Generally, with time and study we iterate from non-discrete to discrete stochastic and eventually to discrete deterministic models as we obtain data to better constrain the models. The eventual reservoir simulation model may be made using both deterministic and stochastic reservoir sectors depending on the distribution of the basic reservoir and fracture data. This merging of sector types is usually performed by a reservoir engineer conversant with the modeling software.

Many of the input parameters listed in Figure 2.3 are not input as individual values, but as parameter population distributions. Previous modeling history can guide the modeler to the proper shape of the distributions (symmetric, log, log-normal, power law, etc.) but these shapes need to be anchored by absolute valves from the individual field being modeled. Examples of parameter distributions are shown in Figure 2.5.

As model inputs, we need the shape of the parameter distribution as well as its' statistical representation, such as values of the most likely, maximum, minimum, and standard deviation. Either we can construct the distribution ourselves from ones that have been applied by others, or we can input statistical parameters of the parameter population and the modeling software will create a logical distribution function.

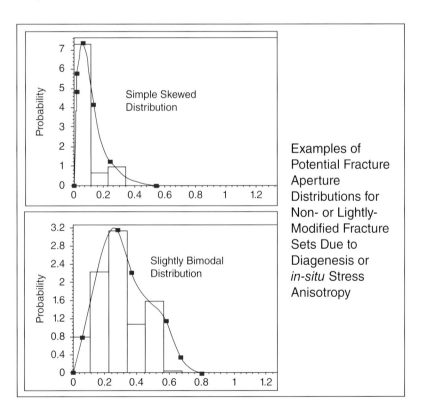

Figure 2.5 This diagram shows two parameter distributions for model input. These are for fracture aperture. Typically, most parameters show a simple skewed, unimodal distribution function (above) while others display a skewed bimodal distribution, or more (below). Few parameters, except fracture orientation, used in modeling display a symmetric distribution.

Depending on data available for the reservoir, we must often rely on different data sources to constrain our distributions. I have developed several approaches to constrain distributions over the years using a variety of geological, petrophysical, rock mechanics, geophysical, reservoir engineering, and drilling and completion engineering approaches. These techniques are applied according to the data types available in the specific field, at that point of study.

As a result, this report will discuss the individual modeling steps and the distributions needed within them and will highlight one or more techniques I have developed along with team members from numerous client teams to constrain those distributions with the data at hand.

Therefore, this report will not tell you the exact process to follow to accomplish completion of the SCFM but rather tell you what needs to be addressed and how that data can be obtained. The process shows how you can have the freedom to generate innovative techniques for data constraint from the data you do have.

3

Fracture Model Creation Workflow

There are certainly many workflows used for creating a Static Conceptual Fracture Model (SCFM). Indeed, each service company and several E&P companies tout their own approach (Bratton et al. 2006; Lange et al. 2009), although many are proprietary. In this book, I utilize a detailed workflow that I have developed over my career. In the early days of work with fractured reservoirs within the industry, approaches were driven by, and often limited to, direct observational data from outcrops or cores. The work was generally non-quantitative and often driven by structural geology and rock petrology approaches. Results of those models were applied dominantly to exploration applications.

However, with time, it became obvious that these reservoirs required multi-disciplinary approaches and some form of fluid flow prediction over time. This along with increasing abilities of simulators to handle complex reservoirs, like fractured reservoirs, led to very quantitative approaches relaying on multi-discipline data integration and quantitative model generation. Output is now utilized across the E&P spectrum from exploration to completion strategies and development of in-fill plans and reserve calculations.

The complete workflow that I use is detailed in Appendix A of this manuscript. A visual schematic of this workflow with timing significance, is shown in Figure 3.1. This figure is similar to the schematic in Figure 1.1 with the addition of data types needed to accomplish creation of both a SCFM and Dynamic Conceptual Fracture Model (DCFM).

An alternate workflow visual that is only on the static side and is in the spirit of the one above is from Bourbiaux et al. (2003), Figure 3.2.

Working through the workflow is not a single-discipline endeavor. It requires work from many technical disciplines; geologists, petrophysicists, geophysicists, reservoir and completions engineers, and rock mechanists. The workflow moves from observational data in all disciplines to integration and interpretation in 3D, eventually leading to a quantitative depiction of the conceptual model. While this workflow is truly a multidisciplinary endeavor, experience shows that there must be a "fracture champion" in the process. This is someone

Static Conceptual Fracture Modeling: Preparing for Simulation and Development,
First Edition. R.A. Nelson.
© 2020 John Wiley & Sons Ltd. Published 2020 by John Wiley & Sons Ltd.

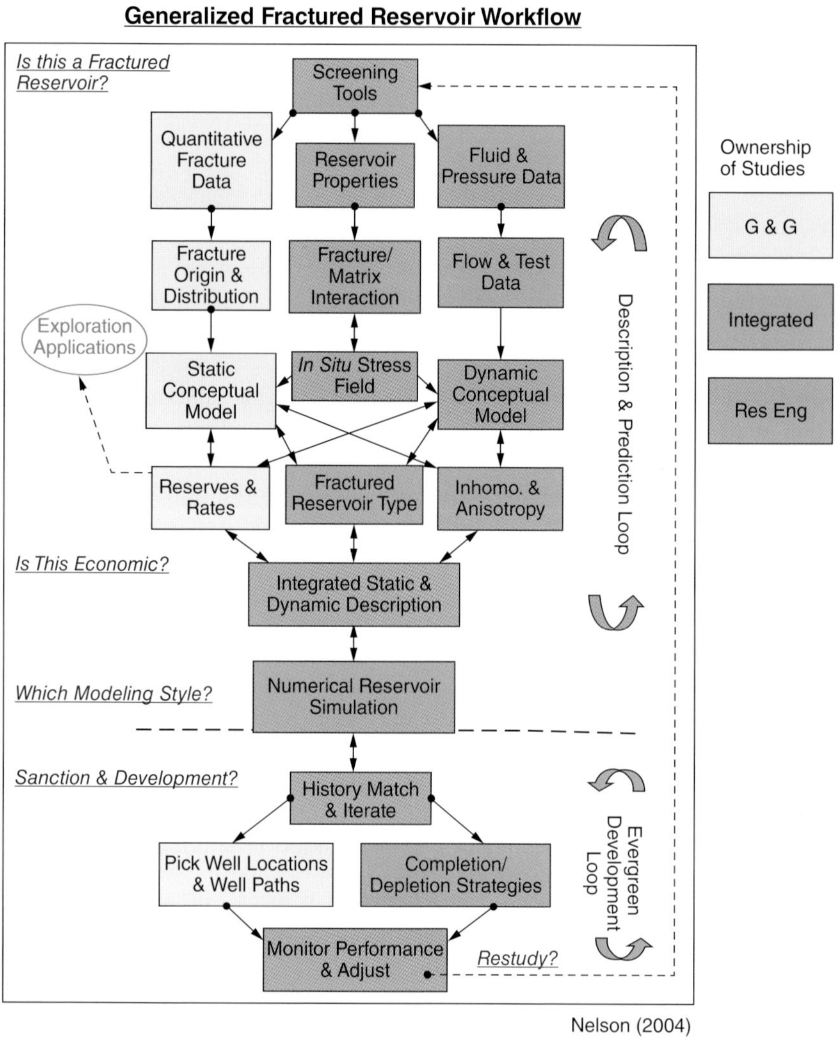

Generalized Fractured Reservoir Workflow

Nelson (2004)

Figure 3.1 A schematic diagram of the modeling workflow for creation of an SCFM. Details of the workflow are presented in Appendix A.

who champions the modeling process and shepherds the work along the way. Important roles of the champion include, (i) getting the necessary work done across disciplines by providing guidance and support for each team member, (ii) fostering communication among the working members of the team throughout the process ensuring that everyone knows where their work fits in the modeling effort, (iii) making sure work output of the disciplines is in the proper format for integration, (iv) facilitating the integration process during team integration

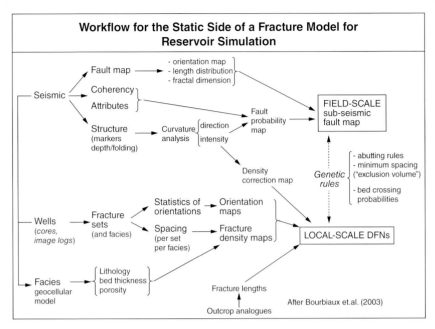

Figure 3.2 An alternative type of workflow for generation of an SCFM from Bourbiaux et al. (2002). This one exclusively covers the static or physical side of the fracture modeling process (*Source:* courtesy of ATC.)

sessions, and (v) act as the project representative to management. Without these roles, work can occur in silos with no one feeling responsible for the final product. These model generation projects need to truly be team efforts.

In my experience, the fracture champion can either be a member of the company management team (supervisor or director) or an external fractured reservoir consultant housed with the team for the duration of the modeling effort. Both approaches work, but which works best depends on the morale and temperament of the team and the management structure. I have worked on both types of studies and find that the external consultant champion often works best, especially within large companies. I think the reason is that the champion is not part of the management structure and team members can work with an external expert and not worry about pleasing the boss and can delve into a technical issue without concerns around any internal company politics of the time. In some studies, the external fracture champion is the person responsible to management in terms of project timing, status reporting and completion.

The remaining sections of this volume will address what is involved in each of these steps or phases in the SCFM creation and will highlight several techniques I have used to quantify the individual parameters from various data sets available in different studies.

4

Gathering Natural Fracture Orientation and Intensity Data Directly

In Figure 2.2, the first four topics listed (in green highlight) involve obtaining data on the orientation, distribution, and physical characteristics of the natural fractures present in the reservoir of interest. This requires observation and interpretations of the reservoir from multiple data sources over a variety of observational scales. This includes relevant outcrops, subsurface core, and geophysical investigations of various types. The following sections will detail many of these data acquisition types.

4.1 Outcrop Based Data

Relevant outcrop studies are perhaps one of the most robust data sets for obtaining quantitative fracture set characteristics. The reason is that outcrops can sample a relatively large volume of potential reservoir rock in 3D. In proper situations, outcrops can allow us to map fracture orientation and fracture intensity along pavements over a relatively large area (individual structure); while simultaneously allowing us to see changes in these parameters in the third dimension or with stacked stratigraphic units. In addition, we can also characterize the morphology of the fracture surfaces themselves (Nelson 2001) although stress unloading at the surface and any surface-related dissolution or cementation can alter the spacing and morphology compared to the subsurface. Only outcrops and core allow for the direct investigation of fracture surfaces directly. In addition, outcrops generally show greater numbers of fractures (although of similar distribution) than their counterparts in the subsurface due to stress relaxation at the surface.

Many exploration and production companies now tend to neglect this important data set often blaming the additional costs of a field program. However, these field acquisition programs are far cheaper than individual 2D seismic lines or even complex Borehole Image Logs (BHI).

Static Conceptual Fracture Modeling: Preparing for Simulation and Development,
First Edition. R.A. Nelson.
© 2020 John Wiley & Sons Ltd. Published 2020 by John Wiley & Sons Ltd.

4.1.1   Requirements for Outcrop Selection

All outcrops are not necessarily useful for constraining fracture characteristics in any specific fractured reservoir or fracture play whether it be "conventional" or "unconventional." Indeed, what is needed are outcrops of the same reservoir unit as we are working within the subsurface, exposed on a similar style structure to what we see in the subsurface. Proximity to our subsurface target is also helpful. Outcrops of different units or rock types in different structural forms, while instructive, cannot be used as direct quantitative analogs for the subsurface.

In addition, the outcrop should include locations that include both pavements (exposed bedding surfaces) and vertical sections sampling multiple beds or units. In this way, we can measure both 2D variability in map form and simultaneously the effect of mechanical stratigraphy in the third direction, Figure 4.1.

Such data is usually acquired manually at the outcrop, which is the best, most detailed way, or by measuring virtual outcrops developed with the use of Lidar imaging, or laser-based imaging of pavements or cliff faces; Geiger and Matthai (2014) and Mah et al. (2011).

4.1.2   Data to Be Collected

Once an appropriate outcrop or series of outcrops has been selected, we need to collect natural fracture data from them. What can be obtained from outcrops, that is unique to all the other data sources, is the size of the fractures; length, height, aspect ratio and fracture set cross-cutting relations.

Stratigraphic Exposure

Pavement Exposure

Figure 4.1   An example of an outcrop that will allow for quantitative natural-fracture analysis from both pavement surfaces (2D analysis) and cross section (1D analysis). For scale the thick green arrow is about 100′ of vertical thickness.

There are many techniques used to collect outcrop fracture data. Examples of such programs include Nelson (1985), Jamison (2016), Gross et al. (1995), Segall and Pollard (1983), Hanks et al. (1995), and McGinnis et al. (2015). Many involve laying out regular-sized areas of pavements and measuring size and orientation of all fractures within that area. Some use box-shaped areas while others use circular-measurement areas. Once an area of measurement is chosen, one approach is to measure each individual fracture within the area. In other approaches, scan lines are constructed within the area and measurements are made on each fracture crossing the scan line. Often two scan lines are drawn with one parallel to structural strike and the other parallel to structural dip; Bosworth et al. (2014), McGinnis et al. (2017), Hencher (2014). Alternatively, three scan lines can be chosen 120° apart. Some define how many fracture sets exist in the measurement area as defined by strike azimuth and lay out one scan line perpendicular to each fracture set. In my work, I have used a variety of scan line approaches, but generally use four scan lines at each potential measurement location including one in the pavement strike direction and one in the pavement dip direction in map view, and in cross section, one parallel to overall bedding and one perpendicular to bedding. The length of the scan lines should be maximized according to the geometry of the outcrop exposure.

The following fracture data should be extracted from the outcrops:

- Measure fracture orientation and its' distribution across the structure in various structural domains; backlimb, forelimb, crest, and hinge(s). This should be done in one stratigraphic unit, if possible. Bedding attitude should be measured at each collection station and the location of the station mapped, preferably by global positioning system (GPS).
- Measure fracture intensity (intercept rate on an oriented-measurement scan line) or fracture spacing (distance between parallel fractures) of individual fracture sets at each measurement station on exposed pavements. This should be performed in one dominant stratigraphic horizon in many spatial stations around the structure.
- In fracture-measurement stations, where it is possible, measure fracture intensity/spacing in multiple stratigraphic/mechanical units in the vertical sense. This will allow for turning vertical intercept rates in wellbores into horizontal fracture spacing for modeling. This is done as the outcrops permit.
- Document any faults on structure in outcrop. Document attitude of the fault, sense of motion, and the width and intensity of the fracture zone surround the fault(s).
- Measure or estimate the percent of open vs filled fractures per azimuth class and what the dominant filling material appears to be. Lump partially open fractures with open fractures. Of course, fracture filling may be quite different from that in subsurface.

- In a couple-measurement stations (high and low potential structural strain positions) record fracture width or mechanical-fracture aperture. These are of course relaxed fractures due to stress unloading but if we take measures in mineral-filled fractures this could indicate possible subsurface width at the time of mineralization. In the best form, taking maximum, minimum and average aperture for each fracture may be appropriate.

Recording the above in a spreadsheet, for example, Excel, allows for easy data consolidation and calculations.

4.1.3   What's Real and Not

As opposed to handling and coring induced fractures in core (see Section 4.2) outcrops often have a different type of induced fractures (not real subsurface fractures) that must be removed from the subsequent natural fracture data set. In general, induced fractures in outcrop, as opposed to core, are related to either unloading of subsurface physical or thermal stresses related to uplift and exposure of the unit or are related to the presence of free or unsupported surfaces such as canyon walls.

In terms of fractures related to free or unsupported surfaces, the classic example is increasing fracture intensity of Mode 1 extension fractures approaching deep river cuts. These topographic-induced fractures usually mimic the geometry of the free surface or canyon wall. In essence, gravity pulls the rock surface toward the center of the canyon, Figure 4.2. The induced fractures are most intense near the cliff surface and die off rapidly in intensity away from the free surface.

Fractures related to unloading of physical or thermal stresses due to uplift and exposure are generally Mode 1 extension fractures that parallel the exposure surface, Figure 4.3.

4.2   Core Based Data

A very important data type for natural fracture interpretation and evaluation is subsurface core. Core observations allow for the determination of real vs coring and core handling induced fractures and the characteristics of the individual fractures including; fracture morphology, inhibition, if any, to fluid cross flow, and fracture intersection angles. Cores exhibit a limited fracture size compared to outcrops and BHI but allow for detailed analysis of the fracture planes and their interaction, more so than other data types. Core also allows for a form of fracture aperture measure as well (unstressed). With core, we can also tie fracture intensity directly to laboratory or log-derived mechanical properties. Even in fracture modeling projects dominated by other data sources, such as BHI and

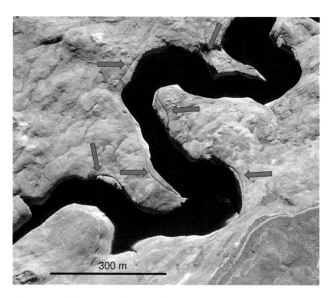

Figure 4.2 The outcrop depicted is Tr-Jr Navajo sandstone along the entrenched just north of the Escalante River near Lake Powell in SE Utah. The arrows depict large-scale induced fractures related to free and unsupported surfaces (cliff face). *Source:* this Google Earth photo courtesy of S. Serra.

Figure 4.3 A granite outcrop with fractures related to a combination of stress relief at the surface and weathering with time. The fractures parallel the free surface of the outcrop or are perpendicular to the maximum extensional stress direction during unloading. The same geometry occurs with rapid thermal cooling of extrusive rocks. *Source:* Google Earth photo courtesy of S. Serra.

seismic attributes, core-based fracture interpretations are very much needed to calibrate or ground truth the larger scale observations.

4.2.1   Types of Core

Core can come in various sizes, lengths and orientations. They can be vertical or horizontal rotary cores or short 1–2-inch horizontal sidewall plugs taken at several positions along the inside of the wellbore.

In terms of continuous core, NX cores are a small-diameter core used in mining applications and stratigraphic reconnaissance, while a 4-inch diameter core is usually taken in E&P applications. There are several formats of core used to quantify the distribution of natural fractures in the industry; whole core, core-slab photos, core slabs, core butts and, core slabs plus core butts laid out together.

Whole-core observation, especially at the rig site, can display important natural fractures. However, it is difficult to see all the natural fractures on the outside of the rough and tooled core surface. Examination of whole core for natural fractures at the drilling rig can be important as handling fracture development is lessened and black-light examination and/or application of a soap solution shortly after the core has been recovered can be very helpful in recording fractures that retain volatile oil or gas seepage that can be unseen weeks or months later when the core is finally analyzed in the laboratory. The bottom line is that whole-core observation of natural fractures is useful for discriminating later handling fractures and important flow planes further along in the process.

Fractured core is often taken with a plastic or aluminum liner to keep it intact during the coring and extraction process. Some operators ship these to the laboratory in complete joints, while other cut the liners into smaller lengths; often 3-foot sections to fit into standard core boxes. The more we cut the core into small lengths, the more difficult it is to piece the core together for possible orientation estimates. A new liner used is the so-called "clam shell," or two overlapping half-moon sections. With this technique, service companies, like CorePro, can piece the core together and, if drilling-induced fractures are present, to orient large sections of the core in 3D space. Even if core is within longer sections of the liner, handling can generate induced fractures. Figure 4.4 depicts a core section being carried by two rig workers generating a large bow to the core often creating additional fractures.

Another important factor in generating handling-induced fractures in horizontal core in particular (often favored in unconventional fractured reservoirs) is the heel of the well or the rapid bend of the well from vertical to horizontal orientation, Figure 4.5. For the horizontal core to be extracted from the well, it must be pulled up through the heel of the well, thus bending the core barrel and core through about 90°. The result is a spaced series of Mode 1 extension fractures across the core. The spacing of these fractures appears to change with the mechanical properties of the layers being cored.

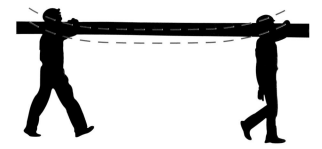

Transporting a full core barrel at the well site can cause
bending of the core and handling induced fractures.

Figure 4.4 Care should be taken when transporting full-core barrels around the rig site as
simple things like carrying the barrel between two rig hands can cause bowing of the barrel
and induced fractures in the core inside the barrel.

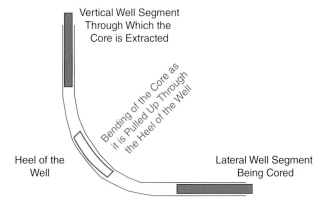

Vertical Well Segment
Through Which the
Core is Extracted

Bending of the Core as
it is Pulled Up Through
the Heel of the Well

Heel of the
Well

Lateral Well Segment
Being Cored

Figure 4.5 Core taken in a horizontal or lateral well must be bent through 90° to be
retrieved from the hole. This creates numerous cross-core breaks that must be interpreted
and removed from the quantitative analysis of the natural fractures present.

The core is shipped from the rig to the core laboratory for slabbing, sampling
and inspection/interpretation. Once core is slabbed, we have two volumes of
rock; usually, the slice along the long axis of the core gives us a 1/3 diameter
core slab and a 2/3 diameter core butt, Figure 4.6. Most often the slabs and
butts are separated and boxed and housed separately. Indeed, some state surveys and foreign countries require either the slabs or butts be sent to them,
thus, housing them in different parts of the country or world.

For fracture screening, we can often start with core-slab photos (often provided by state Geological Surveys or Industry Consortia). Quantitative fracture
intensity can be generated from high-quality core-slab photos and used to high-
grade cores to be observed directly for fracture description. My experience

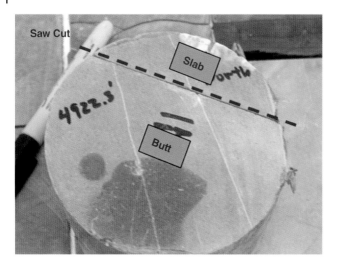

Figure 4.6 The core photo shown from the top depicts the core diameter and the slabbing saw cut. In this case, the slab is ¼ of the diameter while the butt is ¾ of the diameter making the slab even less volumetrically significant compared to the normal case. *Source:* photo is courtesy of J. Lorenz.

from many studies is that interpretation of core-slab photos compared to eventual core slab direct interpretations show that they give a fair representation of relative fracture intensity and distribution. However, they do lose the smallest and/or narrowest fractures.

Fracture examination and quantification on core slabs is very good. In some instances, it is better than whole core as there is a clean, smooth surface to work from. Unfortunately, the slab is a relatively small volume of rock and we can't see fractures that might have occurred only in the butts. This is often made worse as the laboratory workers do not always like putting the saw cut across an open fracture, so they change the saw cut orientation to run parallel to the fracture giving no evidence of the fracture within the slab. For these reasons, I prefer to do my fracture quantification with both the slabs and butts laid out at the same time. I examine and quantify the slabs first and then spot check the butts for any fractures that were missing from the slabs. This also allows a better measure of fracture size and relative orientations.

4.2.2 Data to Be Collected

A variety of natural fracture data can be collected from core analysis. Examples for a non-oriented core include the following;

1) Fracture counts within regularly spaced depth intervals
2) Fracture spacing (average distance between parallel fractures measured perpendicular to the fracture planes)

3) Fracture dip amount (compared to either bedding or core axis)
4) Fracture morphology (open, filled, deformed, etc.)
5) If filled, mineralogy of the filling material or fluid type
6) Fracture height
7) Fracture width or aperture
8) Relevant comments
9) Picture examples of important features.
(All are recorded per sample interval [ft, ½ ft, m, ½ m, etc.])

A decision of which of these attributes to record during observation and the level or form of recording, is dependent on what is to be accomplished in the study or modeling process. Usually, not all the parameters will be described at the level of detail listed above. Therefore, in each fracture study, it is important to design a study-specific data recording sheet. An example of one recording sheet for one study that I used is shown in Figure 4.7. The data recording sheets I use are in spreadsheet format, usually in Excel. Data can be recorded directly into a laptop at the core laboratory or recorded on hardcopy sheets filled out in

Fracture Description from Core (photos, wholecore, slabs, butts)

Well: Name (core top-core base) core length

Depth, ft	Steeply Dipping Fractures Number	Dip, deg	Filling	Shallow Dipping Fractures Number	Dip, deg	Filling	# All Fracs	# O+PO Fracs	Disking Attitude	Pics	Comments/Origin
9,839	7	60 to 90	2, 3	13	0	2	20	4			
9,840	2	60 to 90	2, 3	9	0	2	11	1			Stylolites, s1 vertical
9,841	0			5	0	2	5	0			Stylolites, s1 vertical
9,842	3	60 to 90	3	0			3	3			Stylolites, s1 vertical
9,843	2	60 to 90	3	1	40	3	3	3			Stylolites, s1 vertical
9,844	0			0			0	0			Stylolites, s1 vertical
9,845	1	70	3	0			1	1	H		
9,846	1	90	3	0			1	1	H	x	Induced?
9,847	0			0			0	0	H		
9,848	0			0			0	0	H		
9,849	1	90	3	2	0	2	3	1	H		Induced?
9,850	0			21	0 to 20	2, 4	21	10	H		Filled partings, diagenetic?
9,851	3	90	2	6	0	2	9	0	H		
9,852	5	60 to 90	2, 3	0			5	2	H		
9,853	14	60 to 90	2, 3	1	50	2	15	7	H		
9,854	3	80 to 90	2, 3	0			3	1	H		
9,855	33	60 to 90	2, 4	0			33	15	H	x	2 azi 90 deg apart
9,856	8	60 to 90	2, 4	0			8	4	H		
9,857	0			0			0	0	H		
9,858	8	60 to 90	2, 3	3	10 to 50	2, 3	11	5	H		
9,859	5	70 to 90	1, 2	3	10 to 30	1, 2	8	0			
9,860	5	90	2, 4	0			5	2		x	2 azi 90 deg apart
9,861	6	60 to 90	3, 4	0			6	6		x	
9,862	0			0			0	0			
9,863	0			1	40	3	1	1			
9,864	0			0			0	0			

Tops: Unit 1, Unit 2, Unit 3, Unit 4, Unit 5

Shift legend:
Disking
Up — u
Down — d
Horizontal — h
Variable — v
Mixed — m
Filling
None — 0
Healed — 1
Filled — 2
Open — 3
Part Open — 4

Figure 4.7 An example of an Excel spreadsheet used to describe natural fractures in core. The structure of the spreadsheet and the individual data recorded vary somewhat by the purpose of the project and the condition of the core being described. Stratigraphic units are highlighted in color to make averaging within a unit easier. In this case, the sample interval is 1′, but others can be used. Pics refer to pictures taken at that depth. Number refers to the count of fractures interpreted in that sample height. Similar sheets can be constructed for other input data types, for example, BHI interpretations from the listing files.

pen/pencil during core interpretation and input into the spreadsheet later in the office. As will be discussed later, this data spreadsheet is the basis for further fracture quantification including, Fracture Intensity Curves (FICs) for various fracture orientations, types and morphologies. These can then be correlated with other strip-log data sets, such as rock mechanics and petrophysical distribution logs as well as seismic reflectors and attributes on geophysical cross sections.

4.2.3   What's Real and Not

The most important thing in defining and quantifying fractures in core is determination of "what is a natural fracture and what is not." The ability to do this becomes almost an art form and requires experience. What we observe at the core, is a combination of natural fractures, coring-induced fractures, core-handling induced fractures, and slabbing-related features. Important references for telling real vs induced fractures in core include Kulander et al. (1990), Lorenz (1995, 2008), Lorenz et al. (1990), and Nelson (2001). The most complete publication on the subject to date is Lorenz and Cooper (2018).

With careful foot-by-foot observation of core we can begin to understand what our coring and handling processes give us to sift through to find the real natural fractures for counting. These include;

- Petal fractures (coring)
- Centerline fractures (coring)
- Disking or unloading fractures (coring and extraction)
- Scribe knife (Hugel Knife) breaks (coring)
- Helical torque or twist-off fractures (coring)
- Cross-core breaks (coring and core handling)
- Bedding plane breaks (core handling)
- Core-flexure breaks (core handling)
- Desiccation or drying fractures (storage).

In some cores, the fracture planes observed are dominated by one or more of these induced fractures. There is often interaction between these induced fractures and natural fractures formed before the induced fractures.

In some instances, it is useful to count the various induced fractures along the core. Variations in induced fracture intensity can be used in some cases to highlight subtle changes in mechanical properties in the core related to small changes in rock make-up. In one instance, I observed natural fractures during core being extruded from the core barrels from an Eagle Ford Shale horizontal core that displayed cross-core extension fracture breaks related to flexure as the core was pulled up through the heel of the well. The spacing of the breaks was very regular in both the upper and lower portions of the core but displayed different spacing distance. This was interpreted to be related to small undulations in the well path sampling core from stratigraphically different shale horizons. While, to the naked eye, the shale in the core all looked the same, the spacing of the induced fractures highlighted the mechanical differences.

In a perfect situation, we would like to see the core during removal from the core barrel, in whole-core format prior to slabbing, and post-slabbing slabs and butts. The more views the better, but we work with whatever is available for observation, keeping in mind that the different sources can give slightly different interpretations.

Petal Fractures and Centerline Fractures are an induced fracture assemblage related to the drilling/coring process. As pointed out in Kulander et al. (1990), Lorenz et al. (1990), and Lorenz and Cooper (2018) these fractures are extension fractures related to a high weight-on-string by the driller. These most often occur in hard rocks when the driller increases weight-on-string to maintain penetration rate during coring. The petal fractures originate at the core surface and penetrate downward toward the center of the core, most often in a concave-down geometry. They can occur in pairs on opposite sides of the core and often become periodic with depth in the core. Sometimes the petal pairs across the core link up giving a saddle-like geometry. In many cases, the petal fractures propagate downward to the center of the core and join a centerline fracture that parallels the core axis in a vertical core and splits the core in half. These petal and centerline fractures display mode 1 fracture propagation as extension fractures. Their origin is proven when we observe plumose structure (usually arrest lines) showing a top to bottom propagation direction, Figure 4.8.

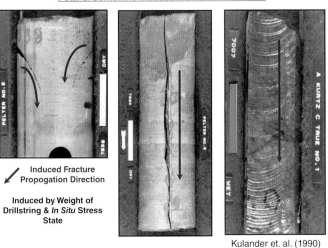

Figure 4.8 Displayed is the geometry and texture of the peta- and centerline-induced fracture assemblage. These are coring-induced fractures and not natural fractures, constructed in the spirit of Kulander et al. (1990).

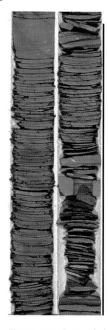

Figure 4.9 Cupped-shaped-up disking fractures in the Eagle Ford Shale. These tend to obscure the natural fractures in these areas of the core.

An alternative view of origin of these drilling-induced fractures, especially in core from horizontal wells, is that they are not due to weight-on-string, but rather high-mud-weight during drilling, N. Nagel (personal communication).

Disking or **Unloading Fractures** are fractures that cut across a vertical core and are closely spaced. They cut all the way across the core and are cup-shaped with the fracture usually turning up at the edge of the core. This indicates that they responded to the core surface while they were propagating. They are unloading fractures formed as the core was extracted from its subsurface stress environment. The rock unloads like an elastic spring parallel to the maximum stress direction in the subsurface, which for relaxed basins is vertical. These occur in some wells and not others, with cores from adjacent wells, sampling the same rock unit, either having or not having disk fractures present. These fractures are probably related to how fast the mud and/or drill string is unloaded in the hole. Highly disked zones make natural fracture interpretation very difficult as they mask other pre-existing natural fractures. Examples of disking fractures in core are shown in Figures 4.9 and 4.10.

Scribe Knife Breaks are induced fractures generated by the scribe knives or Hugel Knives that cut grooves in the outside of the core as it enters the core barrel. The knives were originally introduced to orient the core in 3D with one being the orientation lug. There are three knives and they are set at different angles around the core, Figure 4.11. The one away from the other two is the orienting scribe which is oriented by photographing a compass at the top of the core barrel. This is an archaic approach and is no longer used as comparison of BHI images and corresponding features in core allow the core to be pieced together and oriented.

However, it was found that using the knives in a fractured reservoir helps keep the core intact during the coring process and retard core jamming. The knives, often cutting rather deep grooves in the core surface, sometimes generate coring-induced fractures in the core. When you see an open fracture in the core that emanates from one of the scribes at the core surface, these are potentially induced fractures and not real subsurface natural fractures.

Helical or **Torque Fractures** are coring-induced fractures. They have spiral or helical fracture surfaces and are related to torqueing of the core, much like you get when twisting a piece of blackboard chalk, see Figure 4.12. They are most often seen at the bottom of a coring run when the core is "twisted off" to remove it from the bottom of the hole. An example is shown in Figure 4.13. These are not natural subsurface fractures.

Cross-Core Breaks are induced fractures related to either coring or core handling. For horizontal cores, these fractures are most likely mode 1 extension fractures related to bending when removing the core barrel up through the curved heel of the well. They seem to become periodic (c. 1/ft in high-intensity zones) and change in intensity with rock type or rock mechanical properties. In vertical cores, cross-core breaks with no surface shear texture or mineralization are probably related to core handling, i.e. flexing of the core barrel and liner while extracting and moving the core.

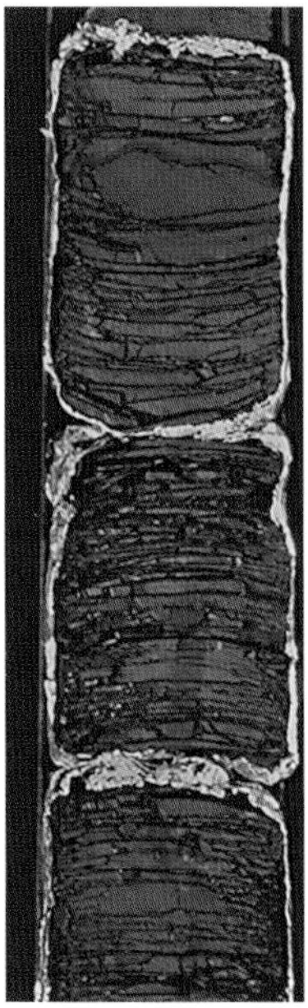

Bedding Plane Breaks are core-handling induced fractures. They are oriented parallel to the reservoir bedding planes and are most prevalent in rocks with strong bedding anisotropy, i.e. shales. These "partings" display either no surface texture or plumose ribs, Nelson (2001). In these rocks, any jostling of the core causes breaks or opening along bedding planes. These are not real subsurface natural fractures.

Figure 4.10 Mixed cupped-shaped up and cupped-shaped down disking fractures in the Jurassic Shale. These tend to obscure the natural fractures in these areas of the core. The question is whether the cup-shaped down is truly a second morphology or have core handlers misplaced a segment of the core in the core boxes upside down.

Desiccation or **Drying Fractures** are handling fractures related to drying of un-cleaned cores. If volatile fluid-filled cores are not cleaned and left for long periods of time the cores can desiccate or dry with time. The result is a polygonal fracture pattern on the outside of the core. These fractures propagate radially toward the center of the core. They are induced by time and drying and are not real natural fractures.

In general, there are rules of thumb for determining real vs induced fractures. They are as follows:

Consider the fractures observed are induced or artificial if;

1) The fracture plane is very irregular or non-planar.
2) The fracture is dissimilar to any other fractures in the core.

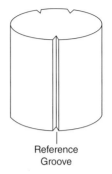

Figure 4.11 A schematic diagram showing the relative placement of the scribes on the surface of the core from the Hugel Knives used to either orient the core or to help piece the core together and keep the core from rotating as it enters the core barrel. *Source: from Nelson et al. (1987), courtesy of American Association of Petroleum Geologists.*

Reference
Groove

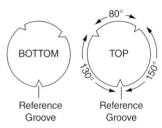

Reference
Groove

Reference
Groove

Figure 4.12 Example of a helical torque-related fracture formed by twisting a piece of blackboard chalk. This is similar to what is seen at twist-off breaking the core from the bottom of the hole. *Source:* photo courtesy of J. Lorenz.

3) The fracture plane emanates from the core edge and dies toward the core center, especially if it dips down into the core.
4) The fracture is horizontal (cross-core) and turns up or down near the edge of the core.

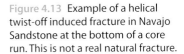

Figure 4.13 Example of a helical twist-off induced fracture in Navajo Sandstone at the bottom of a core run. This is not a real natural fracture.

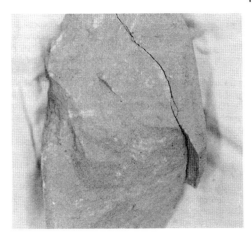

Consider the fractures observed are real natural fractures if;

1) The fractures contain diagenetic cements or minerals not related to drilling of stimulation fluids.
2) The fracture surfaces are slickensided with a consistent inferred movement direction.
3) The fractures are contained entirely within the core, i.e. doubly terminated within the core.
4) The fractures form a reproducible geometry and character with other fractures to form parallel fracture sets.
5) The fracture orientations and character are consistent with inferred fracture origin models (regional, folding, faulting, surface-related).

4.2.4 Quantification

In constructing a Static Conceptual Fracture Model (SCFM) we must generate a quantitative database for elements of the natural fracture system. Therefore, we need to generate a statistical database for fracture occurrence form the core.

When we are defining fractures in core, we are defining the smallest scale of fractures within the reservoir, while BHI and seismic attributes constrain larger-scale fracture elements in the fracture system often missing the smaller-core fractures. This is one reason why core and BHI fracture distributions often do not correlate exactly.

The basic assumptions we make when measuring fractures in core are given in Figure 4.14.

**Assumptions for Natural Fracture Quantification
in Cores**

• Natural fracture follow scaling laws
  – Variation in intensity in core will mimic variations in larger fractures
  un-sampled in core
  – Used to calibrate to fracture in image logs and geophysical attributes, therefore,
  can work out the scaling relations
    • Core stabs
    • Core butts
    • Core photos
    • Resistivity image logs (FMI, OBMI)
    • Acoustic image logs
• Diagenesis & % cemented in the core is the same as for the larger un-sampled
fractures in the rock mass
• Outside of drilling and handling induced fractures, origin(s) of the fractures in the core
are the same as in the rock mass
• Orientations and cross-cutting relations are the same in the core as in the rock mass
• Fracture and mechanical stratigraphy in the core is the same as in the rock mass

---

Figure 4.14 Listed are the assumptions we make when using core observations to quantify
fracture intensity for creation of an SCFM.

Given the fracture descriptions described in Section 4.2.2 above we need to quantify the basic observational data. I take the spreadsheet of fracture description data that has been recorded on a regular sample interval (often 1 ft, 0.5 ft, 1 m, or 0.5 m, depending on the project and data requirements) and quantify the fracture distribution by construction of a boxcar moving average FIC for the core. This requires an averaging window around each individual sample. Most often I use an 11-sample averaging window surrounding an individual sample. However, by changing the sample height and averaging window we can change the shape of the eventual FIC which is a measure of fracture-intercept rate in the measurement direction. By adjusting the sample and window we can better match the shape of wireline logs for the field. However, once chosen, the sample height and averaging window should remain constant for the field or play for consistency.

By creating our FIC, we generate a relatively large database of fracture-intensity measures with depth. As mentioned previously, this curve is a P_{10} measure for modeling. Some workers prefer an alternative display of the data format, that being a Cumulative Fractures Curve. On that curve, the FI rate is represented by the slope of the curve within each unit or layer, Figure 4.15.

This database can also be analyzed statistically, generating parameters such as mean, median, mode, maximum, minimum, standard deviation and

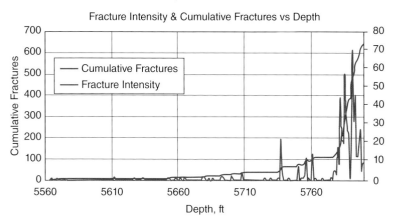

Figure 4.15 Two types of Fracture Intensity Curve (FIC) display along a vertical wellbore. The one in blue is the fracture intensity by 1 ft sample and the other in red is a cumulative fracture count along the wellbore. Both are used by modelers by preference.

the fracture-intensity distribution function. This provides the type of input needed for the fracture-modeling software. An example of the FIC generation and its statistical analysis is given in Figure 4.16 and in different format in Figure 4.17.

In this quantification;

1) **Average Fracture Intensity** is the total number of interpreted fractures divided by the length of core or observational unit.
2) **Mean Fracture Intensity** is the statistical mean of the boxcar moving average fracture-intensity operator.
3) **Fracture Intensity Height** is the summed length of core or observational unit that exceeds a pre-determined intensity cutoff value.
4) **Kurtosis** and **Skewness** are measures of the shape of the fracture-intensity population distribution.

Once we have this statistical database on fracture intensity for all cores or wells calculated in the same way, we can compare the values by well, by unit, by field area and the values can be mapped for comparison to other geophysical and production data (Figures 4.18–4.21).

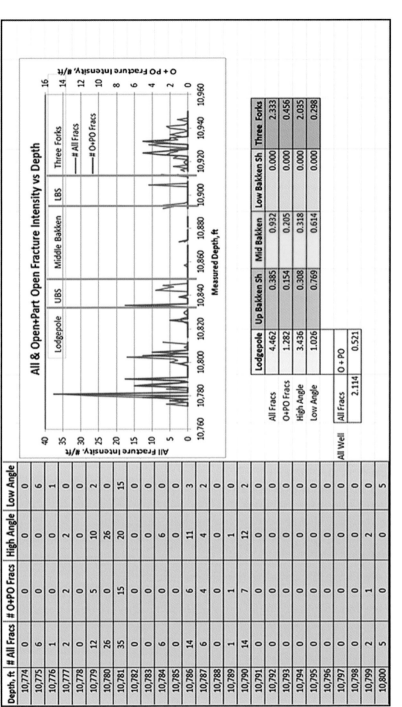

Figure 4.16 An example of a fracture quantification sheet derived from a fracture description data sheet like that shown in Figure 4.7. The quantification is in the form of a natural fracture intercept rate and is broken down for the well and for each stratigraphic and/or mechanical unit. This is then plotted as a Fracture Intensity Curve (FIC) for various elements of the fracture system observed.

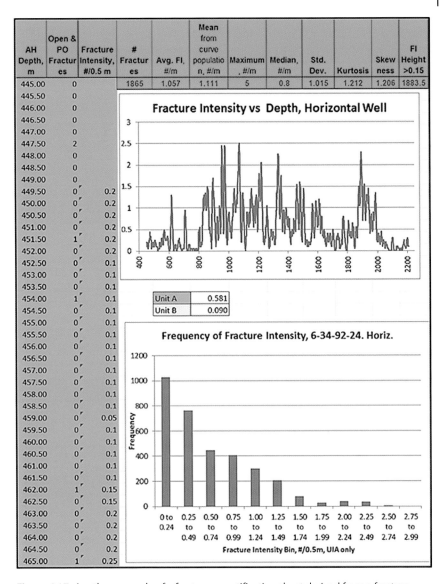

AH Depth, m	Open & PO Fractures	Fracture Intensity, #/0.5 m	# Fractures	Avg. FI, #/m	Mean from curve population, #/m	Maximum, #/m	Median, #/m	Std. Dev.	Kurtosis	Skewness	FI Height >0.15
445.00	0		1865	1.057	1.111	5	0.8	1.015	1.212	1.206	1883.5
445.50	0										
446.00	0										
446.50	0										
447.00	0										
447.50	2										
448.00	0										
448.50	0										
449.00	0										
449.50	0	0.2									
450.00	0	0.2									
450.50	0	0.2									
451.00	0	0.2									
451.50	1	0.2									
452.00	0	0.2									
452.50	0	0.1									
453.00	0	0.1									
453.50	0	0.1									
454.00	1	0.1									
454.50	0	0.1									
455.00	0	0.1									
455.50	0	0.1									
456.00	0	0.1									
456.50	0	0.1									
457.00	0	0.1									
457.50	0	0.1									
458.00	0	0.1									
458.50	0	0.1									
459.00	0	0.05									
459.50	0	0.1									
460.00	0	0.1									
460.50	0	0.1									
461.00	0	0.1									
461.50	0	0.1									
462.00	1	0.15									
462.50	0	0.15									
463.00	0	0.2									
463.50	0	0.2									
464.00	0	0.2									
464.50	0	0.2									
465.00	1	0.25									

Fracture Intensity vs Depth, Horizontal Well

Unit A	0.581
Unit B	0.090

Frequency of Fracture Intensity, 6-34-92-24. Horiz.

Fracture Intensity Bin, #/0.5m, UIA only

Figure 4.17 Another example of a fracture quantification sheet derived from a fracture description data sheet like shown in Figure 4.7. This quantification is in the form of a natural fracture intercept rate and is broken down for the well and for each stratigraphic and/or mechanical unit using a boxcar moving average with inherent population statistics. This is then plotted as a Fracture Intensity Curve (FIC) for various elements of the fracture system observed.

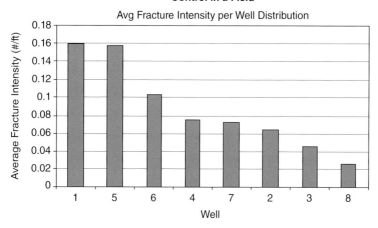

Figure 4.18 Presented is a comparison of fracture intensity by well for a single-field area. In this comparison, we can see a large variation in the natural fracture intensity between wells numbered by age of drilling pointing to the conclusion that some wells should have low-natural fractures effects on fluid flow (e.g. wells 3 and 8) while other wells (e.g. wells 1 and 5) should show much larger fracture effects on flow.

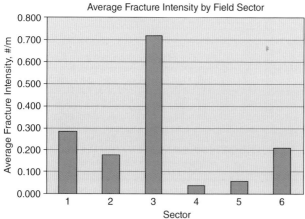

Figure 4.19 Presented is a comparison of fracture intensity by field sector for a large segmented field. In this comparison, we can see a large variation in the natural fracture intensity pointing to the conclusion that some field sectors should have low-natural fracture effects on fluid flow (e.g. sectors 4 and 5) while other sectors (e.g. sector 3) should show much larger fracture effects on flow. This can aid in determining the type of modeling done in each sector and potential development and secondary recovery plans.

Fracture Intensity Can Rank Fracture Effects by Reservoir Layer

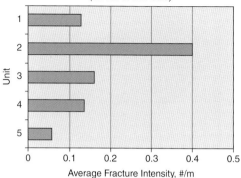

Avg. Fracture Intensity for All Fractures in the Area
(35 Km of BHI data)

Figure 4.20 Presented is a comparison of Average Fracture Intensity by stratigraphic unit for an area of multiple fields. This Fracture Stratigraphy can be the basis for creation of a mechanical stratigraphy. In this comparison, we can see a large variation in the natural fracture intensity pointing to the conclusion that some stratigraphic units should have low-natural fractures effects on fluid flow (e.g. unit 5) while other units (e.g. unit 2) should show much larger fracture effects on flow on a regional basis. The results can aid in determining the degree of flow communication between units of the reservoir. This type of distribution can be cross plotted with average mechanical property variations to create a predictor of fracture intensity from mechanical properties.

Fracture Intensity Height > 0.5

Average Fracture Intensity, 0.5 ft

Mean Fracture Intensity, 0.5 ft

Cogollo Fm

La Luna Fm

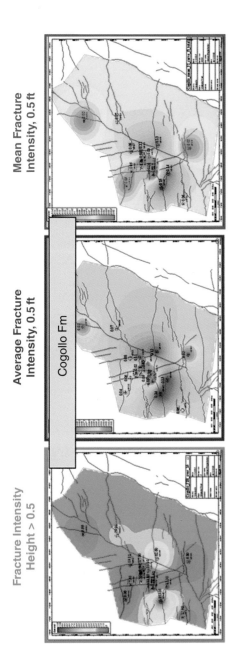
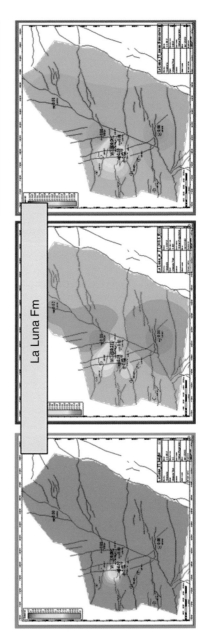

Figure 4.21 Once Fracture Intensity is quantified, the various averaging parameters can be mapped by unit for comparison to other mapped parameters such as petrophysical data, production characteristics and geophysical parameters. Maps of three defined Fracture Intensity calculations; Average Fracture Intensity, Mean Fracture Intensity, and Fracture Intensity Height from two Formations in the Maracaibo Basin of Venezuela.

5

Gathering Natural Fracture Orientation and Intensity Data Indirectly

5.1 Bore Hole Image Log Based Data

Over the last 20–30 years, Bore Hole Image Logs (BHI) have surpassed core and outcrop-based data for the description of natural fractures in the subsurface. An important consequence of this change is a difference in scale of observation of the data types. Indeed, the BHI-based data comes from a larger average size of fractures than what is observed by direct observation. Therefore, care should be taken when using only BHI data to constrain a Static Conceptual Fracture Model (SCFM) for simulation, as the simulation results may give inaccurate (better or worse) predictions of fracture porosity, permeability, aperture, and fluid flow through time. Indeed, our parameter distributions for fracture size, width, and spacing from BHI data will be skewed toward larger fracture-related values.

While it varies from log to log or unit to unit, generally the BHI-based fracture intensities are only about 25% of that seen in core. The quality of BHI images can vary substantially from well to well and run to run as does the quality of the fracture interpretations made from them.

A big difference between traditional core and BHI data is in fracture orientation. We can calculate variations in fracture Intensity in both data sets, but the BHI also directly constrains the orientation of the fractures. In cores, the orientation data is generally not directly available. If needed, we can obtain oriented cores, but these are rarely taken anymore as the process was inaccurate and fraught with errors. Today, we can correlate our core observations with BHI log interpretations over the same lengths of the well and use any drilling-induced fractures (oriented to the stress state) to turn conventional core into a form of oriented core. This does, however, require acquisition of both core and image logs and the hope or a plan to generate drilling-induced fractures to make the correlation.

Typically, exploration and production companies no longer quality control their BHI logs or the fracture interpretations made from them, either on the rig

Static Conceptual Fracture Modeling: Preparing for Simulation and Development,
First Edition. R.A. Nelson.
© 2020 John Wiley & Sons Ltd. Published 2020 by John Wiley & Sons Ltd.

or at the workstation. Most E&P company scientists interact with the log data only from images with interpretations marked on them, sometimes only on paper copies. In rare cases, companies have their own inhouse interpreters and use their own or rented interpretation workstations to carry out the interpretations themselves. In addition, some companies often don't know to request the "feature listing files" from the logging company that records all the individual service company interpretations. These data can be used to quality control (QC) the interpretation and to quantify the data and create FICs to match with other wireline logs and cores.

All interpreters or specialists working with BHI-based fracture data should first QC the images themselves and subsequently the interpretations made from them. Also, all BHI-based fracture interpretations in a field should be performed by the same interpreter or interpreters. Calibration should take place early in the fracture investigation between one or more BHI log types and core taken and interpreted from the same portion of the hole covered by the BHIs. This calibration is critical early on in the study of a field to efficiently integrate multiple data types in the modeling process.

5.1.1 Tool Types and Resolution

There are numerous BHI logging tools used for stratigraphic and structural/ fracture data. Each service company has their own name for their tools and there is a large variation in tool resolution. For natural fractures, there are basically three types of imaging tools; resistivity image logs, acoustic image logs, and visual image logs.

Resistivity Image Logs are the most prevalent for natural fracture interpretation. The tools have many resistivity measurement buttons distributed on eight pads around the wellbore. The data is turned into a resistivity image of the inside of the wellbore displaying a variety of planar bedding and fracture/fault planes and borehole irregularities. Tool names and types include the Formation Micro Imager (FMI) and older Formation MicroScanner (FMS), Oil-Based MicroImager (OBMI), Compact MicroImager (CMI), STAR Imager, and Extended Range MicroImager (XRMI). Output is generally in the form of two image tracks; static and dynamic images. One is the resistivity scale as measured in the wellbore, while the other is the original data recalibrated to a specific length of the log. This tool type has the highest resolution for fracture detection although it varies somewhat between tool types. Borehole coverage varies but none are 100%. The best feature resolution of this tool type is 1/32 in.

Another type is resistivity imaging tool is Resistivity-At-the-Bit (RAB) tool, also called a Logging While Drilling (LWD) tool. One version of this tool type is called a GeoVision Resistivity (GVR) log. The RAB images are the lowest resolution of all the imaging logs making them suitable for finding only the largest of the fractures present, but are of little use in gathering quantitative fracture data.

Acoustic Image Logs image the shape or rugosity of the borehole wall using a centralized spinning acoustic transmitter and receiver pair. This tool type is often used when the well was drilled with oil-based drilling mud making resistivity images of poor quality and in certain cases when hole diameter and/or shape may cause sticking of the resistivity tools. Image output is based on the returning signal recorded in both travel time (shape of the hole) and amplitude (type of material encountered) images. Tool names include Ultrasonic Borehole Imager (UBI), Acoustical TeleViewer (ATV), Circumferential Borehole Imaging Log (CBIL) and Fast Circumferential Acoustic Scanning Tool (FASTCAST). In general, resolution is half that of the resistivity tools, or 1/64 in.

Visual Image Logs are basically downhole television cameras or visual scanners. These tools are limited in depth of usage due to temperature and pressure limitations and have relatively low resolution. They also require the wellbore to be empty or filled with a clear fluid and need a strong down hole light source. These tools help with determining detailed hole conditions but are difficult to use for quantitative fracture analyses. One such tool is called EyeDeal.

Often more than one tool is run in tandem for calibration purposes and response comparison. An example of co-run acoustic and resistivity-based image logs is given in Figure 5.1.

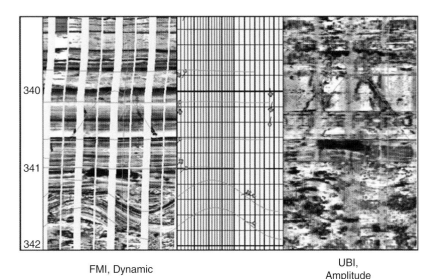

FMI, Dynamic

UBI, Amplitude

Figure 5.1 A log comparison of an Formation MicroImager (FMI) resistivity log (dynamic image) and a corresponding Ultrasonic Borehole Imager (UBI) acoustic image log (amplitude image) in a single well run in tandem. Notice the interpretive difference between the two images.

5.1.2   Data to Be Collected

Typically, interpreted image logs record a variety of planar features. These include:

1) Bedding planes and stratigraphic boundaries
2) Natural fractures with varying filling mineralogy and porosity
3) Faults
4) Drilling-induced fractures
5) Borehole breakouts

For each of these planar features, the following data is recorded:

1) Feature type
2) Depth of the midpoint of the feature
3) Azimuth of the feature plane
4) Dip magnitude of the feature plane
5) A quality of feature pick, usually one to five (only if asked for)

What is generally not recorded, but can be requested, is calculated fracture length or height, fracture intensity or intercept rate and, for the FMI, fracture aperture or mechanical width.

As mentioned previously, the feature interpretation data is recorded in a Feature Listing File that is often in the form of a .las file or an .els file. Once we have these files, we can make orientation plots (Rose Diagrams and Poles to Planes Plots), see Figure 5.2, for all the feature types and Fracture Intensity Curves for various aspects of the fracture data (All, Open, Partially Open, Closed) as well as orientation logs for the various fracture and bedding data. In some situations, the Feature Listing File is not sent to the company and not included in an interpretation report, if one is done. It is often difficult to find these interpretation files within the E&P company requesting the interpretation, or from the service companies running and interpreting the logs. Modeling the fractures is much more difficult if these files are not available for analysis.

Once we determine the orientation of natural fractures for a model grid cell or groups of cells, we need to describe the "orientation dispersion" or the tightness of the orientation data set. Figure 5.3 shows a Rose Diagram depicting the orientation measures for a subsurface data set and the corresponding fracture azimuth distribution diagram. When inputting fracture orientation into the modeling software, Minimum, Most Likely, Maximum, and Standard Deviation of that distribution will be required.

5.1.3   Quantification

The quantification of natural fractures from BHI interpretations is essentially the same as detailed for core analyses (see Section 4.2.4). The biggest difference

All Fractures, A Horizontal Well

Rose Diagram   Pole Plot

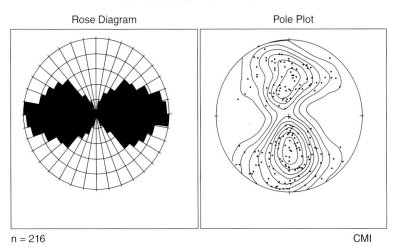

n = 216   CMI

Figure 5.2 Two fracture-orientation diagrams that are typically constructed from the image log interpretations. On the left is a Rose Diagram showing a polar histogram of the azimuth of fracture strike. On the right is a contoured Pole Plot which simultaneously depicts the strike and dip of the interpreted fractures as poles perpendicular to the fracture planes. The orientation diagrams shown are consistent with fractures related to a normal fault state of stress.

(a)   (b)

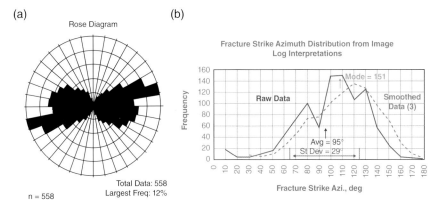

Figure 5.3(a and b) An image log natural-fracture strike data set from a log horizontal well segment in the western USA. (a) A Rose Diagram showing a histogram of strike azimuth in a polar format. (b) is a standard strike distribution plot in two forms. The solid blue histogram is the raw data displayed in 10° increments, while the dashed green histogram is a smoothed data set using a moving average of three data points. The statistical values shown are for the raw data curve. The analyst choosing the data input to the model could choose either curve (or another smoothed curve) to use for data input. The choice between these curves is whether we want to model one fracture orientation or three.

is the fact that quantification from BHI data can be done by fracture azimuth class. In addition, fracture intensities, or the inverse-fracture spacing, for each parallel fracture set of fractures must be corrected for the difference between the azimuth of the fracture set and the azimuth of the deviated or horizontal (parallel to bedding) wellbore. These topics will be discussed later in this book.

It is important to note that when comparing Fracture Intensity from different data sources (outcrop, core, BHI, remote sensing, and seismic attributes) we are comparing quantitative data sets measured at fundamentally different scales, therefore, the numbers will be different at the same location. As will be discussed later, this requires calibration across data sets. However, when combining these different scaled data sets, we can potentially work out the fracture scaling laws for the size of the natural fracture system present.

5.2 Remote Sensing-based Data

Remotely sensed data of the ground surface has been used to infer subsurface fracture and structural orientations and distributions since the beginning of natural fracture analyses. The data is derived either from interpretations of the earth's surface above the reservoirs of interest, and/or in conjunction with the effects of subsurface geology below the reservoir, from breaks and/or faults in the deep crystalline basement. Both approaches define important orientations or disruptions and larger more significant alignments or structural lineation. These orientations, lengths, and spacings once again address the first three parameters listed in Figure 2.2.

5.2.1 Surface Based

5.2.1.1 Various Satellite-based Bands and Altitudes

The original remote sensing work in the industry was done on visual band low-altitude aerial photos, either mosaics of single photos or on stereo photo pairs. Linear and curvilinear features were interpreted and related to geological structure and natural fracture sets. In fact, photo-geology is still done using a distinct set of empirical rules and procedures for performing the work. Later, with the advent of high-altitude aircraft, especially military aircraft, remotely sensed imagery was extended to larger scale images using basically the same set of rules and procedures as low-altitude photos.

In the 1970s, satellite imagery became available for remote sensing interpretation and later from the Space Shuttle. These images were available at a variety of scales and resolution (highest resolution usually restricted to military purposes only). The satellite images could be obtained for a variety of visual and non-visual spectra, including such channels as infrared or thermal IR. Images from these bands of energy can be interpreted alone or in combination by combining bands for interpretation. Figure 5.4 shows an example of natural

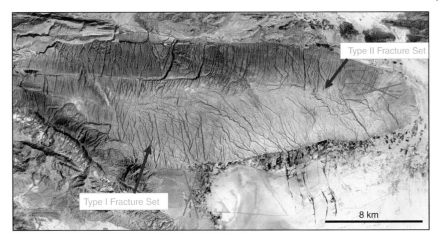

Figure 5.4 A satellite photograph of a large fold in Pakistan. Notice the strong linear fabric to the surface showing a fold-related fracture pattern. If we measured the orientation of the individual fractures and fracture swarms, a consistent pattern would be obvious that would help constrain orientation, length and spacing in a fracture model. *Source:* Google Earth photo courtesy of S. Serra.

fractures on a large well-exposed anticline in Pakistan. The natural fracture orientations related to folding are very evident in the weathering of the surface (Types I and II fracture sets of Stearns 1968).

5.2.1.2 Digital Elevation Models

Today, much has been made of our high-resolution Digital Elevation Models (DEM) from satellite or space station measurements in terms of lineation analyses. Indeed, these data can be displayed at various scales and operated on by various mathematical algorithms to enhance the interpretation. A number of these algorithms attempt to find linear or curvilinear discontinuities in the digital data. Two that are often used are curvature analysis (2D and 3D) and Ant Tracking. Some very successful approaches apply Ant Tracking on a 3D curvature data set derived from the DEM data. These approaches generate objective, quantitative lineation data that can be integrated with other data sets.

5.2.1.3 Lidar Approaches

Lidar is a laser imaging tool. It scans 360° surrounding the instrument (or some more limited angular arc) and details the 3D shape of the surfaces encountered. Lidar is now applied to 3D outcrops for fracture delineation. From the oriented laser and the 3D representation, fracture orientation can be determined directly from the lidar data. In addition, lengths and intensities per azimuth class can be estimated. This data is inherently the same as that of direct outcrop fracture work. The real benefit lies in the digital nature of the data set allowing for easy and rapid quantification of the fracture system(s) present.

In addition, Lidar data can be interpreted remotely when the field person cannot scale the vertical cliff face to make the measurements.

5.2.2 Basement-Based Geophysical Methods (Potential Fields or Gravity and Magnetic Data)

In most structural situations other than thrust belts, structure and tectonics is basically "thick-skinned" or ones whose shallow structure is related to or "forced" from the underlying crystalline rock basement. In fact, even in "thin-skinned" thrust belts, the large-scale geometry is often influenced locally by deeper basement geometry and faulting. Basement geometry and structure can be discerned by deep-focused seismic or gravity and magnetic mapping. Interpretations from relatively cheaply obtained gravity and magnetic mapping are frequently used for defining deep structure. These are often the first data sets utilized in a new exploration area as they exist for most parts of the world and so are readily available and cheap to obtain.

Gravity mapping inherently focuses on variations in rock density, specifically looking at rock type differences at the crystalline basement level. Magnetic mapping inherently focuses on variations in magnetic properties, specifically looking at rock type differences at the crystalline basement level. Linear discontinuities evidenced on these maps are usually interpreted as faults of various type in the basement or perhaps large-scale terrain boundaries, Figures 5.5 and 5.6.

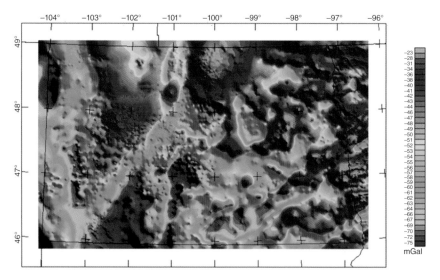

Figure 5.5 Bouger Gravity Map of North Dakota. A large portion of the map includes the Williston Basin. Linear boundaries in the map are interpreted as structurally important Lineation. *Source:* Map is from the USGS website, 2012. Used with the permission of the U.S Geological Survey.

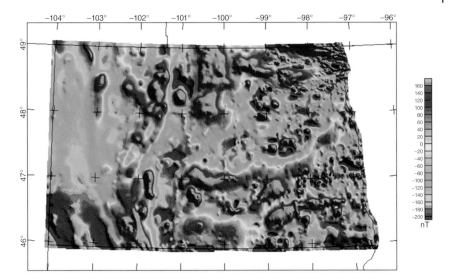

Figure 5.6 Aeromagnetic Map of North Dakota. A large portion of the map includes the Williston Basin. Linear boundaries in the map are interpreted as structurally important Lineation. *Source:* Map is from the USGS website, 2012. Used with the permission of the U.S Geological Survey.

Discontinuities or Lineation from basement potential fields interpretations are integrated with other shallower based data sets, such as surface-based lineation and exploration seismic data.

5.3   3D Seismic Fracture Data Collection

For creation of an SCFM, only 3D and 4D seismic data sets are useful for quantification. In the simplest approach, the surveys are used to develop detailed structural shape maps and sections, from which predictions of fracture orientation and intensity distributions can be made based on empirical models. In another approach, we can use the 3D data to create appropriate seismic attributes used to predict the same type of distributions. Alternatively, in a time of proliferation of hydraulic fracture treatments in horizontal wells, we use passive seismic to listen to natural fracture opening and closing due to earth tides, and monitoring fracture opening signals related to multiple stages of hydraulic fracture treatments.

All these seismic-related data sets should not be used by themselves but must be calibrated with "ground truth" natural fracture distributions and orientations from outcrop, core or well logs. Calibration can come from quantitative data sets or from similarity to empirical analog models. Without such calibration, there may be no quantitative truth to the seismic inferences.

5.3.1 Detailed Structural Geometry

A 3D seismic survey allows for creation of a detailed geometric model of the structures we drill in 3D. These models need to be adjusted by matching to formation tops in wells, or multiple tops in horizontal wells due to the "porpoise" like highs and lows in the horizontal well.

From a well-constrained structural model, we can perform curvature analysis on one or more of the structural surfaces to predict variations in natural fracture intensity. If a formation is relatively brittle and fails by fracturing, areas or volumes of higher curvature should display relatively higher natural-fracture intensity. Higher curvature indicates higher structural strain, and thus higher-fracture intensity, if the rock is deforming by a brittle fracture mechanism. Deformation mechanism can be determined in the rock deformation laboratory under simulated depth conditions, i.e. brittle vs ductile.

An example of structural form and curvature is given in Figure 5.7.

5.3.2 Seismic Attributes

Once the petroleum industry started relying on 3D seismic for both exploration and development, geophysicists started creating seismic attributes to

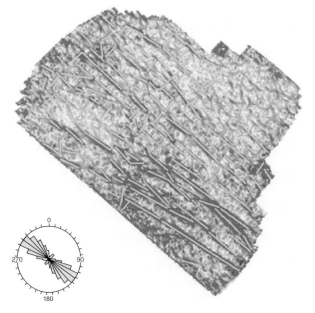

Figure 5.7 A horizon map displaying most-positive curvature along with mapped faults in yellow. The Rose Diagram is the azimuth of the mapped faults. *Source:* Figure from Chopra and Marfurt (2007b), courtesy of Canadian Society of Petroleum Geophysics.

better describe and predict reservoir property distributions. The resulting attribute maps and sections have been used to address variations in relative saturation, porosity, changes due to secondary recovery by water floods (and CO_2 floods), and open natural-fracture distribution, among others. A list of some of the attributes that have been used for natural-fracture distribution (orientation and intensity) is given in Figure 5.8.

A couple of standard attribute examples for natural-fracture prediction are given in Figure 5.9.

The two examples shown in Figure 5.9 both use a form of amplitude mapping, perhaps one of the more popular classes of approaches to date.

Various Seismic Attributes That Have Been Used in the Past to Define Natural Fracture Distributions in the Subsurface

1. Fault & Structural Geometry from Structure Maps & Sections

2. Coherency Attribute Maps

3. Dip Maps

4. Dip Azimuth Attribute Maps

5. Seismic Frequency Maps

6. Seismic Amplitude Maps (RMS, Sum Absolute Amplitude, Volume Amplitude)

7. Interval Velocity Maps

8. Curvature Maps

9. Azimuthal Anisotropy Maps

10. P-impedance & S-impedance Maps

11. Azimuthal AVO

12. Edge Detection Maps

13. Spectral Decomposition

14. 3D Curvature on Coherency

15. Ant Tracking

16. Neural Networks using multiple attributes

17. Seismic Discontinuity Mapping

Figure 5.8   A non-exhaustive list of seismic attributes frequently used to define natural-fracture distributions in the subsurface. Each of these may work in specific situations, but none work in all. "Ground truth calibration" is mandatory to turn the attributes into a local predictor of natural fractures. A description of each of these and how they should be interpreted is given in Appendix B.

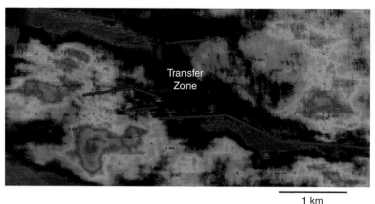

(a)   **Sum of Absolute Amplitudes Over Normal Fault Segments Linked by Relay Zone**

Transfer Zone

1 km

(b)

Amplitude Map
Top Socuy

Lake Maravaibo,
Venezuela

Fault Traces

Stiteler et.al. (1994)

TOP SOCUY
AMPLITUDE MAP
BLOCK I 3D SEISMIC SURVEY

Figure 5.9 (a) A producing area in South America with seismically mapped normal faults displayed in red. The mapped property is the sum of all absolute amplitudes with higher-fracture intensities predicted in areas of low amplitude. Normal faults are down dip to the south, so the pink areas are on the hanging wall. The area between two normal fault tips is a displacement transfer zone; (b) A simple amplitude extraction from a flattened top reservoir section in Venezuela. Here, the lower amplitudes (light areas) are predicted to indicate higher than background fault-related natural-fracture intensity. The largest faults North-North-East (NNE) are left-lateral strike-slip faults. Note the variation in fracture zone width along the faults and where these major faults later cross East-North-East (ENE) faults and fractures and where the faults splay. *Source:* Map is from a presentation by Stiteler et al. (1994).

The concept is that zones of open natural fractures in the subsurface dissipate or disperse incoming seismic waves, thus reducing the amplitude of the returning wave. Frequent applications of amplitude mapping for natural fractures include Root Mean Square (RMS), Sum of Absolute Amplitudes (peaks plus troughs) and simple extractions of amplitude. All are done on a numerically flattened horizon. I have found these simple approaches quite helpful and very straightforward to interpret within the context of structural strain predictions.

All the attribute approaches listed above can give spurious predictions. I think more attention should be paid to those that (i) you understand the physics of, (ii) correlate between multiple attributes, and (iii) fit empirical observations of fracture distribution with respect to structural form; for example, more fractures on the downthrown side of a normal fault, zones of elevated fracture intensity in an envelope 100s of meters around fault surfaces, and increased fracturing in fault displacement transfer zones. In this regard, structural-strain modeling can be important in verifying the attributes. In all cases, attribute maps and sections must be calibrated with real subsurface-fracture intensity or production increases.

5.3.3   Passive Seismic and Hydraulic Fracture Monitoring

Over the last 10 years or so, we have experienced a large increase in hydraulic fracture treatments in horizontal wells. Originally patented by Amoco Production Co. (then Stanolin), hydraulic fracturing was performed for many years in large, single treatments in vertical wells. With the advent of horizontal drilling in unconventional reservoirs (very low permeability rocks like shale), numerous smaller staged treatments were done in the horizontal portion of an unconventional reservoir well; as many as 20–30. Techniques were developed in the industry to monitor the growth and extent of those induced fractures by setting out receiver arrays either at the surface or within adjacent wells, or both, to listen to the breakage of the formation during treatment. The position of each energy release can be pinpointed in 3D and the growth of the hydraulic fracture system can be mapped in 4D, Figure 5.10. From the event clouds during continued injection we can determine both the lateral and vertical extent of the growth of the hydraulic fracture. We can see if the treatment reaches past our acreage and onto another's. We can also see if our hydraulic fractures are "contained" within our reservoir or grow stratigraphically up or down into other units and/or into water-bearing layers.

With experience, it was found that there were seismic events that lay outside the immediate area of the hydraulic treatment and even outside of the formation being stimulated. These can be due to the opening and slip of existing natural fractures in the rock mass that are opened by the elevated fluid pressure of the treatment. These are important as they can impact the success or failure of the stimulation program. The things that can happen when a hydraulic fracture treatment encounters a zone of natural fractures are listed in Figure 5.11.

**Typical Passive Micro-seismic Arrays in Multi-Stage
Hydraulic Fracture Treatment in Horizontal Wells**

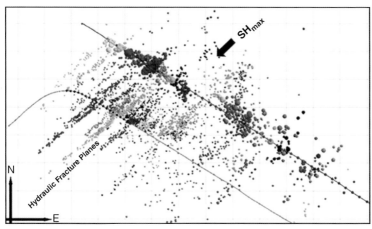

After Chorney et.al. (2016)

Figure 5.10 Two horizontal wells in an Unconventional Reservoir with clouds of micro-seismic events due to multiple staged hydraulic-fracture treatments performed from well heel to well toe. The clustering shows a strong alignment of sources parallel to the maximum horizontal principle stress direction (SH_{max}). Any interference with existing natural fractures will create a more complex source pattern as the existing fractures can divert fracturing fluid and pressure. *Source:* from Chorney et al. (2016) courtesy of Canadian Society of Petroleum Geophysicists Recorder.

**What Natural Fractures Can Do During a Multistage
Hydraulic Fracture Treatment.**

- Divert injection fluid and, thus, retard further hydraulic fracture growth.
- Interact with the hydraulic fracutre and greatly increase fracture/ matrix surface area thus increasing drainage of the matrix and recovery factor.
- Increase vertical propagation of the hydraulic fracture "out-of-zone", leading to erroneous reserve calculation, if not anticipated.
- Open fault zones in the reservoir leading to reservoir seal failure.

Figure 5.11 A listing of the possible consequences of the interaction of existing natural fractures and a propagating hydraulic fracture during a fracture treatment.

Numerous examples exist that show opening mode events can occur up along normal fault planes and their inherent swarm of natural fractures above the perforations where the hydraulic fracture was initiated in both map view (Figure 5.12) and in cross section (Figure 5.13a and b).

**Micro-Seismic Events During
Hydraulic Fracture Treatment**

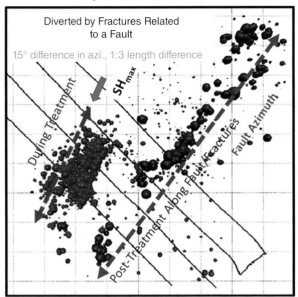

Figure 5.12 Micro-seismic events during and after hydraulic fracture treatment. The red points are events created during pressure breakdown that parallel *in-situ* maximum principal stress direction. The blue points are events created progressively with time after the pressure treatment. These are due to leakage of the pressure along fractures associated with a pre-existing fault. Note the difference in alignment between the red and blue events. *Source:* Kratz et al. (2012), courtesy of CSPG Recorder.

As a follow-up to the fracture treatment approach, researchers started looking at existing fracture opening with respect to ambient subsurface stress changes, or those related to earth tides, Lacazette et al. (2013). Tidal forces related to the pull of the Moon and Sun affect not only the oceans, but the solid earth as well. Tidal displacement of the earth's surface can be as much as 1 m, depending on the relative positions of the Earth, Sun and Moon. As a result, the earth's shallow crust flexes daily in a regular and periodic manner. By passively monitoring the subsurface with seismic receivers, the location of existing swarms of natural fractures and their intensity can be mapped in 3D. These surveys can help in defining the ambient fracture system in the area before the fracture treatment.

It is of paramount importance to understand the distribution (spatial and stratigraphic) of pre-existing fractures in the reservoir prior to fracture stimulation and how those fractures interact with the hydraulic fracture treatment. This interaction can be advantageous by increasing drainage area and reserves, or disadvantageous by causing non-containment of the fracture treatments leading to water encroachment and improper reserve calculations.

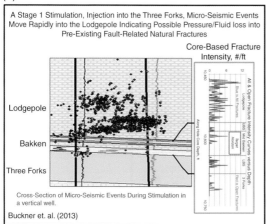

A Stage 1 Stimulation, Injection into the Three Forks, Micro-Seismic Events Move Rapidly into the Lodgepole Indicating Possible Pressure/Fluid loss into Pre-Existing Fault-Related Natural Fractures

Core-Based Fracture Intensity, #/ft

Lodgepole

Bakken

Three Forks

Cross-Section of Micro-Seismic Events During Stimulation in a vertical well.

Buckner et. al. (2013)

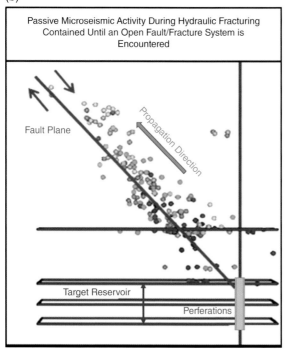

Passive Microseismic Activity During Hydraulic Fracturing Contained Until an Open Fault/Fracture System is Encountered

Fault Plane

Propagation Direction

Target Reservoir

Perforations

Figure 5.13(a and b). (a) Example of non-containment of a hydraulic-fracture treatment in a horizontal wellbore from micro-seismic events. This is from the greater Bakken play and injection was in the Three Forks section below the Bakken. Events rapidly climb stratigraphically, possibly up a fault, past the Bakken (with low G Modulus) and into multiple horizons within the more highly fractured Lodgepole carbonate unit (with relatively high G Modulus). *Source:* after a presentation by Buckner et al. (2013) courtesy of Marathon Oil and (b) Example of micro-seismic emissions during a hydraulic fracture treatment in the Marcellus unconventional reservoir. Events start at the Marcellus target horizon at the wellbore and quickly migrate up a normal fault system approximately 1000 vertical feet, D. Zino Personal Communication.

6

Analyzing the Natural Fracture Data Once Gathered

6.1   Correcting for the Difference Between Measurement Orientation and Fracture Set Intensity

With the advent of horizontal drilling in fractured formations, we found that there is a sampling bias for fracture intercept rate in core or image logs due to the difference between the azimuth of a particular fracture set and the azimuth of the wellbore. The distribution or intensity of a single azimuth fracture set is related to the intercept rate measured perpendicular to the fracture set azimuth. However, the intercept rate we measure in a horizontal Borehole Image Log (BHI) or core of fixed azimuth is usually oblique to the variously oriented fracture sets. Thus, we must correct the fracture spacing (1/fracture intercept rate) geometrically to correct for this difference, Figure 6.1. In this figure, the geometric correction is represented for the quantification of fracture swarms or corridors as well as individual fracture sets within the corridors and in the background.

The correction shown here is a sine function and it is similar in form to that published in Joubert (1998) and Lacazette (1991).

These corrections can be done by grouping fracture azimuths into sets of clustered azimuths and then correcting the spacing or intensity of each azimuth set separately. Alternatively, from BHI log data we can assign a weighting factor for each feature interpreted. This approach is most often cited from Terzaghi (1965).

6.2   Calibration

A major challenge in addressing many of these topic areas is the need to merge data sets measured at different scales in the reservoir and have significantly different inherent resolution. For example, the distribution of fracture sets in 3D in the reservoir can be constrained by several data sets including: small-rock volume measures from core, core slabs and CT or Cat Scans; larger

Static Conceptual Fracture Modeling: Preparing for Simulation and Development,
First Edition. R.A. Nelson.
© 2020 John Wiley & Sons Ltd. Published 2020 by John Wiley & Sons Ltd.

(a)

Assumptions:

Long fracture swarms or corridors with regular properties and all fractures are parallel and have the same permeability.

X = 50 m, y = 200 m, 200 fractures across swarm

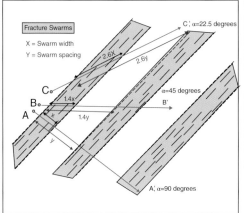

A, x = 50m, y = 200m, 4.0 fractures/m

B, x = 71m, y = 283m, 2.8 frac./m

C, x = 130m, y = 522m, 1.5 frac./m

For a well bore of 1000m:

A = 5 swarms with 1000 fractures

B = 3 swarms with 600 fractures

C = 1 swarms with 200 fractures

(b)

Average Fracture Intensity/Well Correction

Corrected Avg. Fracture Intensity = Measured Avg. Fracture Intensity /sin a

where; a is the angle between well-bore azimuth and swarm azimuth

Swarm Intensity Correction

True Fracture Intensity = Measured Fracture Intensity / sin a

Swarm Width Correction

True Swarm Width = Measured Swarm Width x sin a

Fracture Intensity Height Correction

Corrected FIH = (Measured Swarm Intensity / sin a) x (Measured Swarm   Width x sin a)

Therefore; Corrected FIH = Measured FIH

Nelson (2006)

Figure 6.1 Depicted is a geometric correction needed to correct the relative azimuth between fracture sets and wellbore azimuth. *Source:* from Nelson (2006). (a) presents a schematic of the geometric system for calculation. In this diagram, x is the actual width of the fracture swarm while y is the spacing between swarms. These values change as we move from a wellbore perpendicular to the swarms to more oblique to them. (b) The equations used to geometrically correct the fracture spacings and intensities, and (c) represents the effects on the Fracture Intensity Curve (FIC) using different wellbore approach angles.

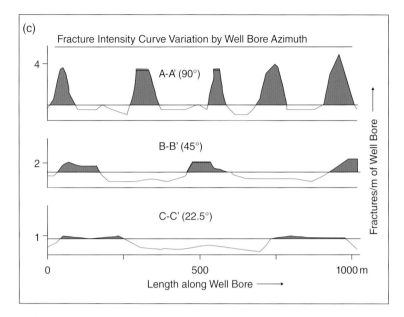

Figure 6.1 (Continued)

volume measures from borehole observations using BHI logs, and large-volume inferences from well tests, production logs, lineation analyses, and geophysical measures from 3D attributes and wellbore shear-wave anisotropy. To effectively accomplish quantification in these important modeling topic areas, we must learn how to calibrate one data set to another and to put all on a common quantitative footing.

Core-based and BHI-based natural fracture interpretations are inherently different in terms of scale of observation and quality of the physical nature of the core and of the images of the borehole. However, the numbers correlate fairly well in Conventional Reservoirs in terms of relative fracture abundance and position in the well. In my experience, BHI interpretations are generally about 50% that of core in these reservoirs, with the highest I have seen at 55%. This is logical in that the smaller fractures evident in core are too small to be imaged in the wellbore. Examples of relatively large differences in natural fracture numbers from core and BHI data sets are found in Wagner et al. (2010), Richard et al. (2017), and Gross et al. (2009), Figure 6.2.

In my work with several Unconventional Reservoirs, the correspondence of fracture numbers between core and BHI counts is poor and significantly lower than in more Conventional Reservoirs. Figure 6.3 shows results for 15 unconventional data set pairs that I have worked with in the past. With one exception, BHI natural-fracture interpretations see only about 0–8% of the fractures interpreted in the same intervals in core from the same zone. Perhaps the reason is that the unconventional reservoirs are dominantly shale which is made

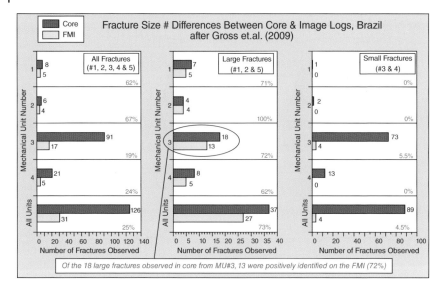

Figure 6.2 Two important things are seen in this diagram after Gross et al. (2009). One is that there are differences in fracture numbers observed in both core and Formation Micro Imager (FMI) between the mechanical units defined in the study. The second is the difference between numbers of small fractures between the two data sets with the small fractures predominantly seen in the core relative to the image logs. *Source:* courtesy of M. Gross.

Natural Fracture Intensities in Corresponding Pairs of
Core & BHI in Several Unconventional Reservoirs

* Reservoir 1 – 2 vertical wells where *BHI is only 0.8% of core.*
* Reservoir 2 – 3 wells w/ mixed results.
  * 1 horizontal well w/good results where *BHI is 42% of core.*
  * 2 vertical wells w/poor results where *BHI is nearly 0% of core.*
* Reservoir 3 – 2 wells w/ poor results.
  * 1 vertical well where *BHI is 4% of core.*
  * 1 vertical well where *BHI is 200% higher than core.*
* Reservoir 4 – 1 vertical wells w/ poor results where *BHI is 0.6% of core.*
* Reservoir 5 – 1 vertical wells w/ poor results where *BHI is 2% of core.*

"Borehole Image Logs, while good at defining natural fractures in conventional reservoirs, have difficulty imaging natural fractures in unconenventional reservoir sections."

Figure 6.3 Above is a list of success rates from comparison of natural-fracture interpretations based on both cores and Borehole Image Logs (BHI) over the same wellbore intervals. These are all from unconventional reservoir sections. What is expressed is the percentage of natural fractures from core observations interpretable on the Image Logs. While results in conventional reservoirs can be in the 50–60% range, those for unconventional shale reservoirs are generally much less than 10%.

up of very small grain size and is very laminated and thus more difficult to image in the image logs. Clearly, we must change how we acquire the BHI data in these reservoirs or place a greater emphasis on core for our modeling efforts.

However, most projects that require creation of a Static Conceptual Fracture Model (SCFM) have a mix of available data types to constrain fracture intensity (FI). This is due to data strategy changes from exploration to development, to secondary recovery, drilling strategy changes during the life of the drilling program, and a general lack of focus on fracture description early in the life of the project. An example of one such project is shown in Figure 6.4. In this project, the data sources varied from fracture descriptions in core, fracture interpretations from Formation Micro Imager (FMI) BHI, and fracture interpretations from a resistivity-at-the-bit (RAB) imager log (GeoVision Resistivity or GVR log). In Figure 6.4, the mean FI (vertical intercept rate) calculated by a boxcar moving average of fracture occurrence for all units in one structural area for 12 wells is plotted in histogram format from highest to lowest FI. This distribution of wells shows that the higher FI wells are represented by core observation measurements, while the FMI-based wells represent the middle intensities, and the GVR wells dominate the lowest intensities. This distribution is logical as it depicts the range in resolution of the data sets. Interestingly, this spread in calibration numbers was later made worse with the change in strategy to move to horizontal wellbores, where there was concern in using the FMI tool due to hole size and bending at the heel of the well. The new data source chosen was the UBI

Mean Fracture Intensity for All Fractures Varies with Data Type

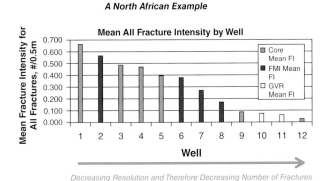

Figure 6.4 A North African example of fracture intensity (FI) distribution among wells using different data sources. Notice that because of the differences in resolution the core-based wells generally exhibit the higher fracture intensities while the Formation Micro Imager (FMI) based wells display the mid-level fracture intensities and the GeoVision Resistivity (GVR) borehole image log (BHI) at the bit display most of the lowest fracture intensities. *Source:* after Nelson (2011b), courtesy of the Gulf Coast Association of Petroleum Geologists.

log (Ultrasonic Borehole Imager) which has a resolution of one-half that of the FMI log. At the time of compilation, no well contained more than one data set type for calibration. We are left with a combined data set for FI that varies from well to well and by resolution of the tool acquiring the numbers, and no data to calibrate one to another. Uncertainty in the model is, therefore, relatively high.

This shows the need for at least one well to contain multiple data types from either multiple runs of different logs or coupling more than one tool together during a logging run. The bottom line is that we must have calibration points if trying to merge different data sets to constrain a modeling parameter.

Another calibration example shows the differences in FI between multiple logging runs and fracture interpretations. Figure 6.5 shows an unusual situation where fracture data was available from two horizontal wells; an original and a parallel sidetrack. The reservoir is a Conventional Fractured Reservoir. One horizontal wellbore had a horizontal core that was oriented using bed dip and interpreted for natural fractures. Both horizontal wellbores had FMI logs with bedding and fracture interpretations. All three data sets were interpreted by different service companies. Analysis of the data sets shows three interesting results:

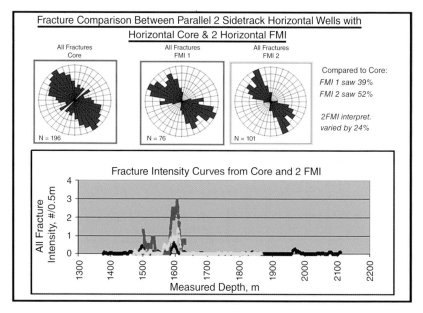

Figure 6.5 An example of a two-well constraint on fracture intensity (FI) in a Conventional Fractured Reservoir. The data is from a horizontal well and a horizontal sidetrack well parallel to the first. Formation Micro Imager (FMI) data on fractures was available from both wellbores. In addition, core was available from one of the horizontal wells. All three data sets depict a similar fracture orientation on the Pole Plots of fracture strike above and location of higher-intensity fracture swarms as shown by the Fracture Intensity Curves below. What is different, is the fracture numbers and fracture intensities between data sets with the core displaying much higher FI in general and within the common fracture swarms. *Source:* after Nelson (2011b), courtesy of the Gulf Coast Association of Petroleum Geologists.

1) Fracture orientation, as depicted in the Rose Diagrams of fracture azimuth are quite similar and well within interpreter variability.
2) The three FI curves constructed for the three data sets, created from a box-car moving average approach using an identical averaging technique, show remarkable similarity in highlighting the position of higher FI zones within the wellbores; and
3) The calculated horizontal FI depicted in the peak fracture swarm or corridor and in the data set, as a whole, are very different in magnitude between the data sets. Indeed, the FMI data sets display only 39 and 52% of the fracture numbers compared to the core interpretation over the same interval, with 24% variability between the two FMI interpretations.

Results of this analysis allow for a quantitative calibration between the core-based and FMI-based fracture intensities that could be used field wide. While the results can seem dramatic, one should remember that, in general, there are differences in resolution between the core and FMI data sets, with the core showing much smaller fractures than can be interpreted in the image log. Some argue that the FMI scale of observation is actually the more important one for fluid flow as many of the smaller fractures may not be connected or contribute significantly to flow.

On a larger scale, we can see the effect of data type between large numbers of wells. Figure 6.6 is a histogram of vertical FI measured for 130 vertical wells from eight producing regions of the world covering a variety of structural and stratigraphic situations. FI is a fracture intercept rate, expressed here as the number of fractures observed per meter (#/m) calculated using a boxcar

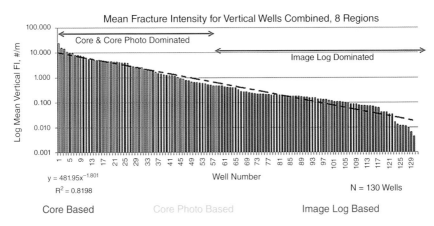

Figure 6.6 Depicted in this figure is vertical fracture intensity (FI) distribution calculated in the same manner for 130 vertical wells in eight productive regions of the world as measured by this author. Also shown is the data type used to calculate the fracture FI in that well (core, core-slab photo, and borehole image (BHI) logs (Formation Micro Imager [FMI], Ultrasonic Borehole Imager [UBI], etc.). The higher 45% of the wells are dominated by core and core photo-based intensities, while the lower 55% of wells are represented by BHI wells. *Source:* after Nelson (2011b), Courtesy of the Gulf Coast Association of Petroleum Geologists.

moving average technique of similar averaging parameters (Nelson 2004). FI is plotted on a log scale and the distribution is fitted with a power-law function. The FI in the wells displayed was derived from three source types; core, core-slab photos, and BHI. We can see in this figure that the upper 45% of the FI wells are dominated by core and core photo-based measurements, while the lower 55% of the FI wells are almost exclusively borehole image interpretations. Of course, no discrimination by mechanical properties, stratigraphic break-down, or structural situation was made in constructing this diagram.

Similar results can be seen for a compilation of horizontal wells with hori-zontal FI or horizontal fracture-intercept rate in Figure 6.7. In this figure, we can see a similar trend in the FI distribution seen in Figure 6.6. However, in the horizontal wells calculated in the same way as the vertical wells, the core-based wells are represented in only the top 12% of wells displayed, while the BHI wells dominate the lower 88% of wells compared.

The bottom line is that the different data sets used to quantify FI give inher-ently different numbers of fractures. This is a problem in properly modeling the quantitative fracture distribution in fracture models for simulation. Similar differences can be seen between the different BHI types and with correlation of all to outcrop measurements (not addressed in this book).

The distributions of FI measurements in large data sets show that quantita-tive assessment for creation of an SCFM must consider where the data comes from. To effectively model fractured reservoirs we must be able to calibrate the

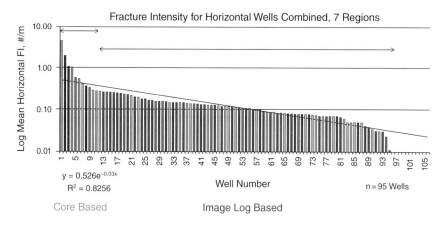

Figure 6.7 Depicted in this figure is horizontal fracture intensity (FI) distribution calculated in the same manner for 95 horizontal wells in seven productive regions of the world as measured by this author. Also shown, is the data type used to calculate the FI in that well (core and borehole image [BHI] logs (Formation Micro Imager [FMI], Ultrasonic Borehole Imager [UBI], etc.). The higher 12% of the wells contain a large portion of the wells measured by core and core photo-based data sets, while the lower 88% of wells are represented by BHI wells. *Source:* after Nelson (2011b), courtesy of the Gulf Coast Association of Petroleum Geologists.

differing data sets one-to-another. This is usually completely lacking in many projects or the attempts to calibrate the data are inadequate.

I have experienced numerous calibration attempts that did not work. A list of some of these is as follows:

1) Different BHIs types in different wells
2) BHI in one well and core in another well
3) Core in a vertical pilot and BHI in the horizontal wellbore
4) One BHI in a vertical pilot and a different BHI type in the horizontal wellbore
5) Outcrop and core or BHI in different parts of the section
6) Outcrop and core or BHI in different structures and/or structural types.

These attempts are obviously inadequate because they do not investigate the same rocks in the same boreholes (or horizontal and vertical pairs of wellbores) or do so with different tools or data types. Resulting data is poorly calibrated or unconstrained and make for a poor SCFM and, therefore, a poor predictor of future flow behavior.

Proper calibration of data sets for quantification of FI requires a focused data-acquisition approach started early in the history of the project. Recommended data acquisition for proper calibration includes the following:

1) A well with all BHI types that will be used in the field over time, run in one hole at the same time
2) Core taken over part of the interval covered by the different BHI's
3) Core in a vertical pilot and core in the horizontal wellbore
4) BHI in a vertical pilot and the same BHI type in the horizontal wellbore to construct vertical to horizontal FI conversion
5) Outcrop and core/BHI from the same section in very similar structural types and shape near the field of interest.

These data acquired early in an exploration or development project allow for quantitative calibration of the various data sets needed to quantify FI in an SCFM. This is done by deriving fracture multipliers when moving from one data set to another, regardless of which set is chosen to best characterize the reservoir model.

Also, as stated earlier, the various geophysical data sets **must be** calibrated with subsurface measurements of fracture system properties from core or BHI interpretations to be predictive.

6.3   Determining Natural Fracture Origin from Fracture Distributions and Morphology

A detailed discussion of the various types or origins of natural fractures can be found in Nelson (1985, 2001). The types described in those references are included in Figure 6.8.

- **Tectonic Fractures**
  - *Fold-related, Fault-related*
- **Regional Fractures**
  - *Joints, Cleat*
- **Contractional Fractures**
  - *Chickenwire, Diagenesis-related, Columnar Joints*
- **Surface-related & Induced Fractures**
  - *Weathering, Spall, Unloading, Diagenesis related*

Figure 6.8 This figure shows the basic types of naturally occurring fracture systems in rocks as described in Nelson (1985, 2001).

In Nelson (2001), it is pointed out that it is imperative to determine the origin(s) of natural fractures in outcrops and in the subsurface to predict fracture distributions away from control. In practice, this is done by interpreting the strikes and dips of observed fracture sets and their variation, both laterally and vertically in the well or the outcrop. From the fracture measurements, we can construct both Rose Diagrams (polar histograms of fracture strike) and Pole Plots (stereonet depictions of poles-to-planes showing the simultaneous strike and dip of fractures, and when contoured the azimuth dispersion of the fracture sets). In the subsurface, we typically create these plots from interpreted BHI data (or measurements in oriented core) from the Feature Listing Files, see Section 5.1.2. By interpreting these plots along with any indicators of slip along the fracture surfaces, plumose markings, and/or paleo-stress direction indicators from stylolites we can, in many cases, constrain the probable origin or origins of the natural fractures present. The following will show real-life subsurface distributions used to constrain fracture origin.

Regional Fractures have the simplest strike and dip patterns in outcrop or wellbores. As discussed in Nelson (1985), regional fractures tend to be perpendicular to bedding and can occur as a single set of parallel fractures, or two fracture sets perpendicular in azimuth to one another, that is orthogonal fractures perpendicular to bedding and perpendicular to one another. If they display surface marking on the fracture surfaces, they will generally display "plumose markings" indicating origin as Mode I fracture opening or as tensile or extension fractures, Figure 6.9.

A Rose Diagram (polar histogram of fracture strike) and Pole Plot (strike and dip simultaneously displayed in 3D) template for regional orthogonal fractures is given in Figure 6.10a and b. This is an example of regional orthogonal fractures in North Africa as interpreted in the subsurface. A schematic diagram showing the general geometry of these fracture sets is given in Figure 6.11 where we see two sets of fractures that are steeply dipping and perpendicular in azimuth to one another, after Nelson (2010c).

For a discussion of Systematic and Non-Systematic fracture sets as well as the equivalent Face Cleat and Butt Cleat in coal see Nelson (2001).

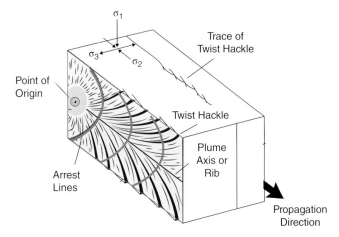

Figure 6.9 A schematic showing the surface marking of a Plumose texture indicating a Mode I (opening mode) crack or an origin as an extension or tension fracture forming parallel to the maximum and intermediate stress directions and perpendicular to the minimum stress direction at the time of formation, diagram generalized from Kulander et al. (1990).

(a)

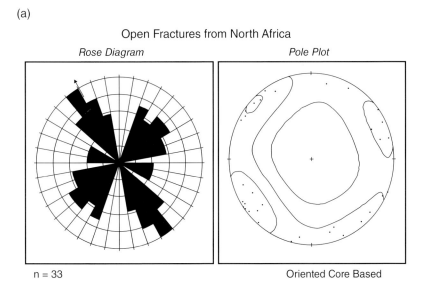

Figure 6.10(a and b) The two diagrams are data representing a system of regional orthogonal fractures as depicted in, (a) as a real Rose Diagram of fracture strike azimuth on the left and accompanying countoured pole plot on the right, and (b) as a schematic Pole Plot showing the vertical and orthogonal nature of the strike and dip of the fractures simultaneously.

(b)

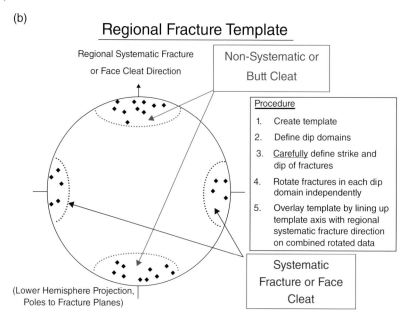

Figure 6.10 (Continued)

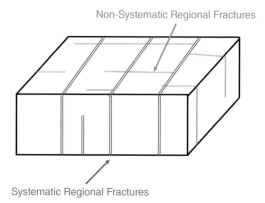

Figure 6.11 A 3D schematic diagram representing regional fractures in a reservoir. The first formed, more through going set is termed the systematic regional fracture set; while the second formed set frequently abuts the first formed set and is generally less well developed and is termed the non-systematic regional fracture set.

Tectonic Fault-related Fractures are the next simplest fracture system to interpret from fracture orientation and distribution data. In general, we should see three fracture orientations related to a fault regardless of the type of fault. We should see two conjugate shear fractures and a bisecting extension fracture related to the fault displaying the state of stress causing the initial fault, Nelson (2001). All will be parallel in strike to the fault surface. One shear fracture will parallel the fault plane and the other shear fracture will be conjugate to the first, or dipping 50–60° to it. The extension fracture should bisect the two shear-fracture planes. The actual orientation of these fractures sets is defined by the orientation of the fault surface, which in turn, is defined by the type of fault or the orientation of the stresses at the time of faulting, see Figure 6.12.

An example of fracture distributions dominated by fractures associated with a normal fault is shown in Figure 6.13a and b.

An example of fracture distributions related to a possible strike-slip fault is given in Figure 6.14. Here, shear and extension fractures are vertical but vary in strike from parallel to the potential fault, conjugate to the fault, and bisecting the two. Evidence of slickensides or slip on the shear fractures should be horizontal for pure strike slip. For large wrench faults, secondary folds and faults related to deformation of the affected fault zone can complicate the observed natural fracture distribution.

Tectonic fold-related fractures are much more complex in distribution than the above and display fractures related to beam-bending moments in 3D. Pole plots of a complete fold-related system are made up of three dominant conjugate fracture patterns (two shear fractures and a bisecting extension fracture) showing causative paleo maximum stress directions in different orientations, Nelson (2001).

As defined by Stearns (1968), and shown in Figure 6.15, conjugate sets 1 and 2 are interpreted on bedding surfaces and display a paleo maximum stress direction lying parallel to bedding for Set 1 down the dip direction of the fold and Set 2 in the strike direction of the fold. Set 3 is defined in cross section and displays

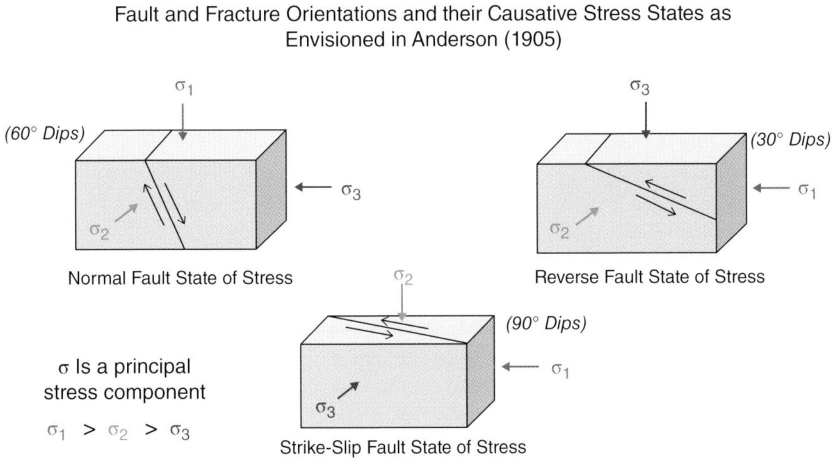

Fault and Fracture Orientations and their Causative Stress States as Envisioned in Anderson (1905)

Figure 6.12 Schematic diagram derived from the concepts of Anderson (1905) on "The Dynamics of Faulting" depicting the relation between stress state at the time of faulting and the orientation of the potential fault planes and, therefore, the orientation of fractures associated with and surrounding the fault surfaces.

(a)

Filled Fractures Related to a Normal Fault

Strike Diagram   *Pole Plot*

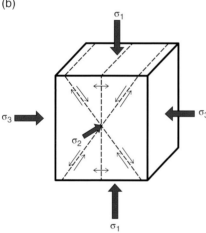

n = 83   FMI Based

(b)

Figure 6.13(a and b) (a) Left is a Rose Diagram format of natural fractures interpreted from a Formation Micro Imager (FMI) log with an accompanying Pole Plot on the right depicting a dominance of fractures associated with a normal fault state of stress, possibly associated with a normal fault striking NNE with opposing 60° dip and possible vertical fractures bisecting them. The West-North-West (WNW) trend is secondary and perhaps related to another fracture set. (b) is a schematic diagram showing the geometry displayed. *Source:* after Nelson (2010c).

the interpreted paleo maximum stress direction either perpendicular to bedding (Set 3a) or parallel to bedding (Set 3b). These are all probably related to stretching Nelson (2001) or extension parallel to strike and parallel to dip and bending of the structural beam, if using a co-axial strain model for distribution.

Filled Fractures in a North African Gas Field

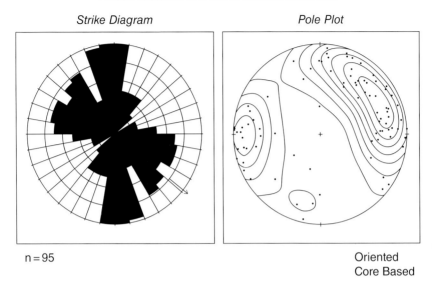

Strike Diagram

Pole Plot

n = 95

Oriented
Core Based

Figure 6.14 These natural fractures depicted in a Rose Diagram on the left and a Pole Plot on the right were interpreted from an Formation Micro Imager (FMI) log depicting the dominance of fractures associated with a strike-slip fault state of stress, possibly associated with a normal fault striking North-North-West (NNW) with vertical conjugate shear fractures with vertical extension fractures bisecting them. *Source:* After Nelson (2010c).

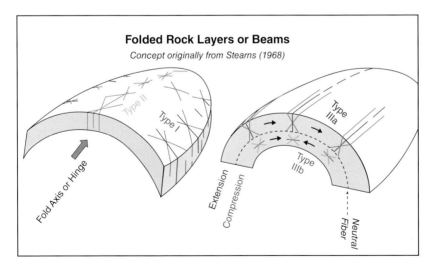

Folded Rock Layers or Beams
Concept originally from Stearns (1968)

Figure 6.15 A generalization of the four different fold-related conjugate fracture sets originally defined in Stearns (1968). All four sets are distributed around the fold but Sets 2 and 3 are most prolific in the fold-hinge zones or areas of rapid change in dip. *Source:* from Waterhouse et al. (1987).

As defined in Stearns (1968) the fold-related fracture sets are inferred from assumed paleo stress directions at the time of folding or during the folding process. However, we can define them by strain at time of fracturing as well. A conjugate fracture system shows shortening in what Stearns (1968) calls the maximum stress direction and extension in what is called the minimum principle stress direction. Therefore, we could define the fracture sets by strain patterns during folding. If the reservoir rock is deforming by brittle fracture, ductile flow of the rock is not possible leaving a mixture of brittle conjugate fracture sets to accomplish strain in 3D. The result is a mixture of fracture sets depicting extension in the dip direction (Set 2) and concurrent extension in the strike direction (Set 1) and in cross section due to bending moments (Set 3).

The result of these mixed conjugate fracture sets is a pole plot with much greater complexity than other fracture origins with 17 potentially developed individual pole clusters, Figure 6.16. However, depending on the subtilties of the strain pattern of the fold, a smaller subset or orientations may be dominant.

Other fracture types such as Contractional and Surface-Related Fractures, are not well defined by orientation plots. They are better defined by distribution with respect to bedding, paleo exposure surfaces, diagenetic events and thermal stresses. See Nelson (2001) for a more complete description of those types.

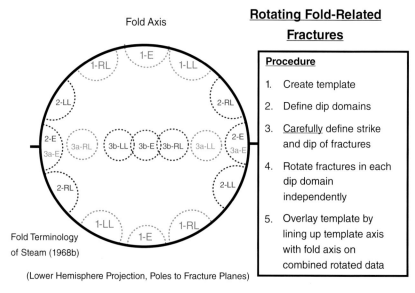

Figure 6.16 A schematic natural fracture Pole Plot showing the potential locations of poles to planes for the fracture sets described in Figure 6.15. E denotes extension fractures, while RL and LL denote right-lateral shear and left-lateral shear fracture positions. The diagram assumes horizontal bedding. To best see the positions accurately, all measurements should be rotated to bedding horizontal. *Source:* After Nelson (2011a).

6.4 Mapping Natural-fracture Orientation and Intensity

Once we have compiled the data for natural-fracture orientation and intensity from our various data sources, the next step is to look for spatial variation in these properties in the scale of our project (regional scale, field scale, model/ simulation scale). This involves mapping the parameter values in 2D and 3D. By mapping in various ways, we can compare our fracture orientations and intensities or clusters to other structural, mechanical, petrophysical, geophysical, *in-situ* stress, and production data sets. In Figure 6.17, a field-scale fracture

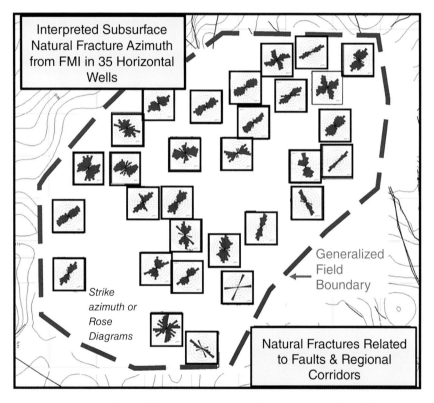

Figure 6.17 Interpreted natural-fracture orientation in the main portion of a Middle Eastern oil field mapped on the structure map for the main reservoir. This rich data set comes from 29 interpreted Formation Micro Imager (FMI) logs in horizontal wells. All logs were interpreted by two interpreters. There is a strong dominance of NorthEast-SouthWest (NE–SW) trends in the data and these parallel the Systematic Regional Fracture direction and fracture corridors related to sub-seismic faults in the area. Additional NW–SE trends parallel the seismically mapped faults in the field. It is interesting that prior to fracture mapping the main fracture trend in the reservoir was considered to be NW–SE, or parallel to mapped faults. *Source:* after Nelson (2010b).

orientation map displays the 2D variation in natural-fracture orientation for the main reservoir unit in a Middle East Field. These fracture-orientation Rose Diagrams were derived from FMI interpretations, all done by two interpreters in one service company regional field office.

Maps such as the above show us the trends of natural fractures and areas of overprinting of trends. This variation in orientation can be used to constrain a static fracture model leading to creation of a Discrete Fracture Network (DFN) for reservoir flow simulation. Typically, the orientation Rose Diagrams are loaded as images in a 3D visualization package (like Petrel) for display at the take point of the reservoir in the well. The parameter distributions of orientation are loaded separately.

The various forms of calculated FI (Average Fracture Intensity [AFI], Mean Fracture Intensity [MFI], Fracture Intensity Height [FIH], etc.) can be mapped in a similar fashion to orientation, although digitally instead of images. In the same 3D visualization software, we can load the FI values at the appropriate top of the reservoir or adjacent units, as well as for averages for each of the reservoir sub-units or facies. Figure 6.18 shows one such FI map for the same Middle Eastern field in the last figure.

As discussed previously (Section 4.2), we can quantify several types of FI relevant to matching to other data types. Three of these are AFI, MFI, FIH and

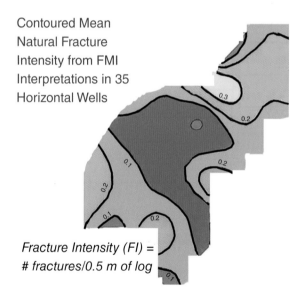

Contoured Mean Natural Fracture Intensity from FMI Interpretations in 35 Horizontal Wells

Fracture Intensity (FI) =
 # fractures/0.5 m of log

Figure 6.18 The trends of Average Natural Fracture Intensity over the same general area as shown in the fracture orientation map of Figure 6.17 for a middle eastern oil field. Here, we see low FI in the central portion of the field (dark green) and higher FI in the northern and southern portions of the field. This variation in intensity is directly related to reservoir facies and environment of deposition that controls the mechanical property variation within the field and, therefore, the Fracture Intensity (FI). *Source:* after Nelson (2010b).

Application of Various Fracture Intensity Parameters

- AFI & MFI relates well to bulk production properties.

  Should be used in comparing wells within and between areas, in comparing different reservoirs and when correlating with cumulative production and EUR.

- FIH relates to well rates.

  Should be used when correlating with Kh, PI, max or initial rate, water-cut & wellbore skin.

- FIC helps define where fluids exit and enter the wellbore

  Should be used when correlating with mud losses, PLTs, and the position and properties of faults.

Figure 6.19 The various Fracture Intensity (FI) or distribution values are listed along with the data types that they are intended to correlate with best.

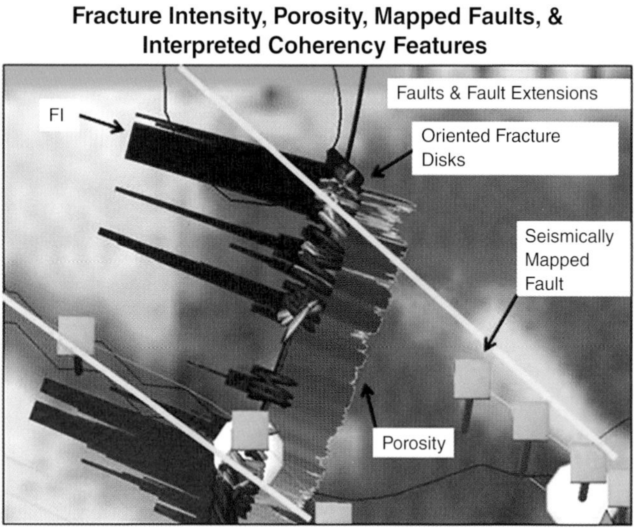

Fracture Intensity, Porosity, Mapped Faults, & Interpreted Coherency Features

Figure 6.20 The Petrel map depicts several data sets in 3D associated with a horizontal well in a field in the Middle East. This map view shows the horizontal wellbore, an All FI curve along the wellbore, oriented fracture disks color coded by open or filled morphology, and a wireline porosity curve. These are depicted on top of a coherency map with a thick yellow fault interpretation from the coherency along with the pillars or matchsticks representing seismically mapped faults. The thin yellow lines represent the position of faults and fault extensions based on integration of all the data. *Source:* from Nelson (2004).

the Fracture Intensity Curve (FIC). Each correlates with different other data sets, Figure 6.19.

In addition to mapping fracture orientation and various FI representations by unit and sub-unit, we can also load the individual fracture strike and dip and FI variations along the wellbore in 3D. This enables us to match to other wireline log suites, geophysical interpretations, structural interpretations, and production logging. The data can be loaded as FICs, fracture strike logs and fracture dip logs along the wellbore for visualization. Fracture attitude displays can also be created by loading the fracture data for oriented disk presentation with the disks displaying the position, strike, and dip of each interpreted feature type (open and closed natural fractures, bedding, faults, drilling-induced fractures, borehole breakouts) color coded by feature type. An example of such a 3D display is given in Figure 6.20. These integration displays in 3D become critical during Team Integration Meetings.

This type of display shows the value of data integration in 3D in one of the various 3D visualization packages. All Technical Team members involved can see their data in the context of others on the team.

7

Gathering and Analyzing Structural Data

7.1 Structural Surface Maps and Sections

A key to working with subsurface fractures and in static fracture modeling is having good quality, well-constrained, structural maps for the horizons of interest. This allows for quality interpretations of geological structures, such as folds and faults, and the location of significant structural discontinuities such as fracture corridors. Generally, structural maps based solely on 2D seismic control are not of sufficient detail for fracture prediction.

7.2 Analysis of Structural Surfaces

In general, the primary application for modeling we get from well-constrained structural maps and digital surfaces is in the form of local shape changes of the surfaces, particularly those that have a linear trend to them. These can be broken down into discontinuity analysis and lineation analysis.

7.2.1 Discontinuity Analysis

Linear discontinuities in structural surfaces are often associated with areas of increased fracture intensity or fracture corridors.

7.2.1.1 2D and 3D Surface Curvature in Depth

The most frequently utilized technique for defining and predicting important natural fracture-related discontinuities in structural surfaces is surface curvature. If the rock of interest is strong and brittle, high curvature can equate to high strain that is then dissipated by creating more fracture surface area resulting in relative high-fracture intensity. Curvature from a structural surface is calculated from the second derivative of the curve of the structural surface. It is basically

Static Conceptual Fracture Modeling: Preparing for Simulation and Development,
First Edition. R.A. Nelson.
© 2020 John Wiley & Sons Ltd. Published 2020 by John Wiley & Sons Ltd.

the rate of change in dip in 2D. The curvature can be expressed in a variety of forms with directional or mean significance. However, all require that the structural surface is constructed in depth and not seismic time. Examples include:

1) Dip curvature
2) Strike curvature
3) Mean curvature
4) Positive curvature
5) Negative curvature
6) Maximum curvature
7) Minimum curvature
8) Contour curvature
9) Shape index
10) Curvedness.

Examples of the application of curvature in subsurface mapping for practical purposes are given in Hennings et al. (2000) and Watkins et al. (2018).

In addition, surface curvature can be calculated simultaneously in 3D and this is called Gaussian Curvature, Mynatt et al. (2007). All give different representations of shape changes in the structural surfaces.

Caveats to applying curvature analyses to predicting fracture intensity are included in Figure 7.1.

The most frequent use of structural surface curvature is to find flexures or fold hinges in the data. These hinges are expected to contain high-intensity

---

Caveats for Curvature Application

- Curvature is assumed to predict structural strain; more curvature = more strain = more fracturing
- Assumes all strain goes into brittle fracturing.
- If strain partitioning takes place (competing deformation mechanisms) curvature will not be accurate to predict just fracturing.
- A poorly constrained structural map or surface will result in questionable curvature maps.
- Curvature quantitatively highlights structural hinge zones; therefore, it predicts higher intensity fractures parallel to the curvature/hinge zones.
- There are both primary hinges (best displayed by dip curvature) & secondary hinges (best displayed by strike curvature or a mix of strike and dip curvature).
- The curvature versus fracture intensity plot appears to be non-linear with relatively low curvature values showing a relatively flat relation to intensity, thus not a quantitatively significant predictor of fracture intensity at low curvature.
- At higher curvature, the curvature/FI relation is steeper and a better predictor of FI. The shape of the relation should be worked out empirically for the unit and locality of interest.

Figure 7.1 This list presents caveats or things to consider before interpreting curvature attribute maps of a horizon for natural fracture prediction and input into a Static Conceptual Fracture Model (SCFM).

Figure 7.2 Schematic structural contours map segments displaying the definition of Primary and Secondary Hinges that surface curvature attempts to illuminate. Primary Hinges parallel structural contours and are evidenced by abrupt changes in contour spacing. Secondary Hinges obliquely cross structural contours or are a locus of points where the structural contours spread out along a line. These often occur toward the plunge of a fold. *Source:* from Nelson (2011a).

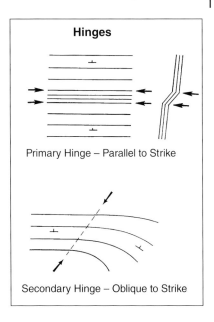

fracture swarms striking sub-parallel to the hinge. If the hinge is parallel to the structural contours on a fold this is called a Primary Hinge and is well defined by Dip Curvature. If the hinge or wrinkle in the structural surface crosses the structural contours this is called a Secondary Hinge and is adequately defined by strike curvature, Figure 7.2. Thus, when creating a numerical static fracture model for simulation, we can use the Dip and Strike Curvature maps to model Primary and Secondary Hinges and the orientation and intensity of their internal fractures differently.

An example of a co-rendered dip and strike curvature map in a folded area is shown in Figures 7.3 and 7.4. In this Figure pair, we can see high-curvature lines paralleling the structural contours (Primary Hinges) and high-curvature lines oblique to the structural contours (Secondary Hinges).

Depth-surface curvature maps can be used as input to seismic attribute discontinuity mapping such as Ant Tracking (of Schlumberger).

7.2.1.2 3D Time Curvature

An alternative approach to treating structural surfaces for fracture prediction is utilizing a seismic time surface rather than a depth surface. As such, it is really a seismic attribute, most of which will be handled in a later section of this manuscript; Appendix B. In the 3D time curvature approach, the time surface is investigated in the 3D-time volume of the 3D survey. Curvature is calculated at each pixel or nodal point in the volume by averaging it with its' 15 neighbor pixels in 3D. Examples of the approach are given in Chopra and Marfurt

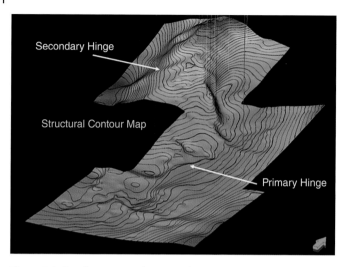

Figure 7.3 A well constrained structural contour map in an area where the reservoir top surface is gently folded. Reds are structurally high, and blues are structurally low.

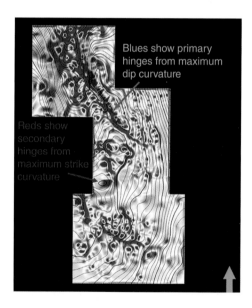

Figure 7.4 A co-rendered dip max curvature and strike max structural curvature map for the same area covered in the Figure 7.3. Blue colors are high-max dip curvature and reds are high-max strike curvature. This map displays both Primary and Secondary Hinges for static fracture modeling.

(2007a). Some prefer this approach to surface-depth curvature as it does not need to be corrected for velocity profiles that, if poorly constrained, can introduce depth-surface errors, and therefore introduce curvature errors. An example of a 3D-time curvature map is given in Figure 7.5.

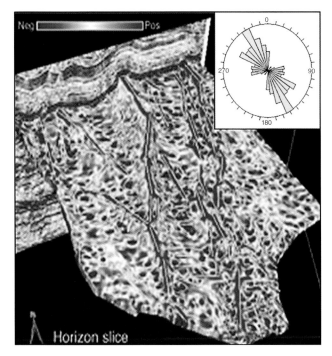

Figure 7.5 A 3D time maximum curvature map from Chopra (2009), with mapped faults in yellow and a corresponding Rose Diagram of fault strike azimuth. *Source:* courtesy of First Break.

3D-time curvature maps or data sets can be used as input to seismic attribute discontinuity mapping such as Ant Tracking (of Schlumberger).

7.2.2 Lineation Analysis

One of the oldest approaches to predicting sub-surface structure (folding, faulting, discontinuities, and natural fractures) is surface lineation analysis. The concept is that linear or curvilinear features interpreted at the earth's surface or at the level of the structural basement can indicate structural features at the reservoir level of interest, Prost (1994) and Papadaki et al. (2011). This approach has been applied at the surface using various data sources such as aerial photos (single and stereo pairs), topographic maps, satellite photos, and most recently Digital Elevation Models based on satellite data, Gabrielsen et al. (2018). At the basement level, these analyses were performed on seismic surveys and gravity and magnetic maps. These analyses were applied heavily prior to the application of 3D seismic surveys.

As an interpreted lineation appears to be validated geologically over time by workers in an area, the individual lineation features are often named for communication purposes. A classic example of this is the relatively weakly deformed Williston Basin in Montana and the Dakotas, Figure 7.6.

With time and substantial amounts of quality seismic surveys and wells, these named lineation are further defined as to fault type, if appropriate, and zones of increased natural fracture intensity and/or local control of deposition and diagenesis.

7.2.2.1 Surface Lineation

Linear or curvilinear discontinuities interpreted from various data sets used to describe the morphology of the earth's surface have proven useful in predicting sub-surface structural geology. The general techniques used in such interpretation are detailed in Nelson (1983).

The various data sets that have been interpreted for surface-based lineation include the following;

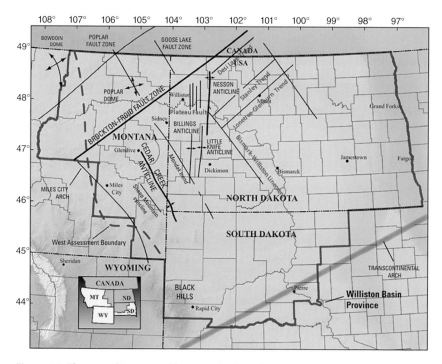

Figure 7.6 The named interpreted lineation for the Williston Basin. *Source:* modified from Gerhart et al. (1990) plotted on a base map from Anna et al. (2010). Several of these have been validated over the years as wrench faults with the aid of 3D seismic and horizontal wells.

- Topographic maps (direct or shaded)
- Aerial photos
- Satellite and near space-based photos
- Digital Elevation Models (DEMs).

In the interpretation, a mix of classic photo-geology and "trendology" is applied. The interpreter looks at various scales of the data set looking for predominantly straight alignments in the data. These may indicate dominant fault system trends of various types, fold axes and, cross-strike discontinuities (CSD) such as in thrust terrain or fold belts. Historically, interpretation was done manually, thus introducing an element of operator variability. However, today we have access to tools that can act on digital data sets to highlight linear features in the data. One such application is Ant Tracking^{TM} (Schlumberger) generally used to detect potential faults directly from forms of seismic data. Alternatively, we can apply curvature analysis to the DEM data set to highlight linear features in the data.

The interpreted lineation data is utilized for both position and fabric patterns of features. For position we attempt to use the actual position of the individual linear as significant to the position of features in the subsurface. An alternate interpretation style is to use the combined fabric of the lineation data in terms of orientation and length as an indicator of dominant structural fabric in the area for stochastic modeling, Nelson (1983).

7.2.2.2 Basement Lineation

An alternate approach to lineation analysis is to investigate the rock package not at the earth's surface or at the top of the package, but at the base of the sedimentary rock package, or usually at the top of the crystalline "basement." This technique is especially appropriate for areas dominated by thick-skinned tectonics; even though surface lineation analyses are appropriate for both thick-skinned and thin-skinned tectonics. In general, our reservoir of interest is sandwiched between the surface and basement lineation.

Basement lineation are generally derived from seismic data imaged at the basement level or from potential fields data and maps. These data seek to find important fault trends and basement terrain boundaries both of which can affect the geology of the overlying section over time and multiple periods of deformation. Linear features that are evidenced on both surface and basement data sets are most likely to affect our reservoir. Examples of basement of lineation from both seismic and potential fields data for a portion of the Bakken Play in North Dakota are given in Figures 7.7 and 7.8.

7.2.2.3 Structural Timing and Sequencing

Multiple periods of structural development in an area generate multiple fracture sets displaying different properties and different interpreted tectonic

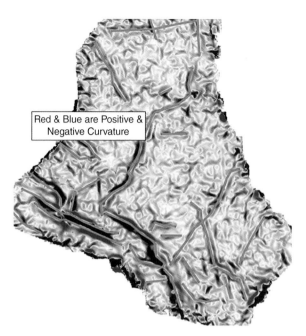

Red & Blue are Positive &
Negative Curvature

Figure 7.7 A map of Maximum Curvature operated on a seismic structural surface from just above the Pre-Cambrian basement in a portion of the Bakken Play. Blues and reds indicate positive and negative curvature, respectively. In green are the interpreted basement lineation which separate the two. Inference is that these features would affect structure in the overlying rock section over time.

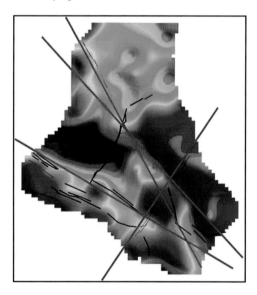

Figure 7.8 A Bouguer Gravity map for a portion of the Bakken Play shown in Figure 7.7 with seismic-based basement lineation overlaid in black and gravity lineation interpretations overlaid in blue. Note the very good correlation in several of the interpreted features. *Source:* courtesy of USGS.

Periods of Interpreted Folding & Faulting in Onshore Kuwait

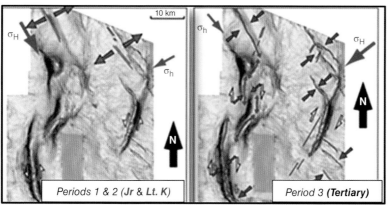

Periods 1 & 2 (*Jr* & *Lt. K*)   Period 3 *(Tertiary)*

After Richard et.al. (2017)

Figure 7.9 Structural sequencing in Central Kuwait as interpreted in Richard et al. (2017). The sequence is used to forward model natural-fracture development using a mechanics approach in the creation of an SCFM for the area. *Source:* modified from Richard et al. (2017), Courtesy of Society of Petroleum Engineers.

stress states. The interpreted sequence of structural formation and natural fracturing can be used to forward model natural-fracture development in the modeling process. An excellent example of this process is detailed in Richard et al. (2017). An example from that paper is given in Figure 7.9.

8

Gathering Constraints on Fracture Aperture

In creating a quantitative fracture model for simulation, one parameter that is very important to quantify is the aperture or width of the fractures. It is also the most difficult to constrain. The importance of the aperture is that it controls both fracture porosity and fracture permeability in flow equations and in the simulator. These parameters are both a function of fracture aperture and fracture spacing or the average distance between sets of parallel fractures, Figure 8.1.

Natural-fracture aperture distribution patterns in the subsurface can be distinctly bimodal (or more) due to preferential cementation and diagenesis, as well as, due to azimuthal *in-situ* stress variation, Figure 8.2.

Fracture aperture is highly stress dependent, where high *in-situ* normal stress components acting perpendicular to the fracture planes reduce the aperture and, therefore, the fracture porosity and permeability. In fact, small reductions in aperture will have a relatively small change in fracture porosity but a relatively large negative change in fracture permeability, as aperture is a simple function in the fracture porosity calculation but a cubed function in the fracture permeability calculation, see Figure 8.1.

We can constrain fracture aperture from several sources including outcrops, core, borehole image logs, laboratory experiments and well-constrained reservoir flow tests. In fracture porosity and permeability calculations and simulations fracture aperture is hydraulic aperture (fluid flow measure) rather than mechanical or kinematic aperture (physical measure). The two are very different measures and are related as shown in Barton (2007) and Hooker et al. (2014). In the literature over the years these two terms often tend to be used interchangeably.

Inherently, the various techniques available to us to constrain fracture aperture are measured under very different stress conditions; unstressed, under partially stressed reservoir conditions, and fully stressed to reservoir conditions, Figure 8.3. These give progressively smaller apertures with stress.

Static Conceptual Fracture Modeling: Preparing for Simulation and Development,
First Edition. R.A. Nelson.
© 2020 John Wiley & Sons Ltd. Published 2020 by John Wiley & Sons Ltd.

Fracture Porosity & Fracture Permeability are a
Function of Fracture Aperture & Spacing

$$\Phi_f \% = (e / D + e) \times 100$$

$$K_f = e^3/12D \times \rho g/\mu$$

(For 1 set of parallel fractures)

e - aperture
D - spacing
ρ- Density
g - Accel due to gravity
μ - Viscosity

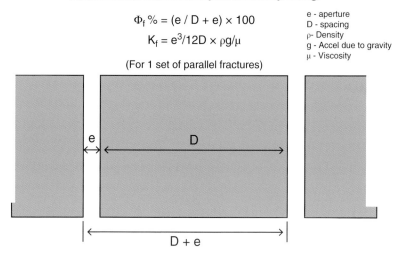

Figure 8.1 A visual representation of fracture aperture in the calculation of fracture porosity and permeability. This representation is for only one set of parallel fractures. The equations are for hydraulic aperture and not mechanical aperture. Also, the fracture aperture will vary by the fracture set orientation within the *in-situ* stress filed.

These varying measurement conditions result in a mix of hydraulic and mechanical aperture measures.

8.1 Unstressed

Unstressed natural-fracture aperture measurements can come from three sources; outcrop, core, and CT scans of core. All give a measure of unstressed mechanical aperture. In outcrop we can use calipers, width comparators (card of lines of varying constrained width, Ortega et al. (2006), or in some cases spark plug gap measurement sets (Marrett 1997). Of course, in outcrop, fractures can be mineralized (simply or complexly) or open (original or influenced by dissolution) with diagenesis related to the subsurface or due to weathering at the earth's subsurface. As such, the data set is considered a relatively poor representation of the subsurface apertures. However, the change in aperture by azimuth of the fracture sets in outcrop can be important in constraining a Discrete Fracture Network (DFN) for simulation through the flow anisotropy it causes. Therefore, we may be able to constrain percent flow anisotropy even if the absolute values of fracture porosity and permeability may contain significant error.

In analysis of natural fracture apertures in core, we still measure apertures in an unstressed environment; however, all diagenetic alterations seen are

Example of Potential Bimodal Fracture Aperture Distributions for Natural Fracture Sets that have been Altered by Dissolution &/or Cementation from Single or Multiple Periods of Diagenesis or *in-situ* Stress Anisotropy.

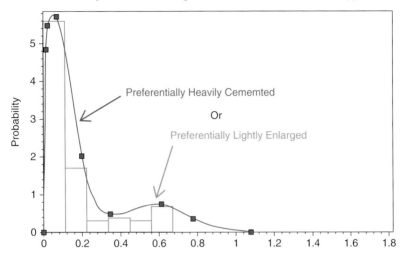

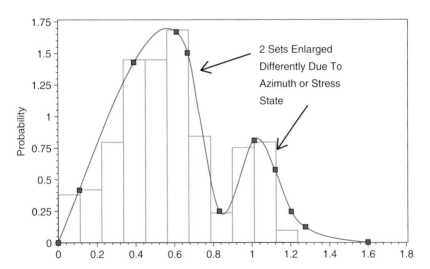

Figure 8.2 Bimodal distributions of fracture aperture for data input into a Static Conceptual Fracture Model (SCFM). The bimodal nature of the distribution can be caused by preferential diagenesis of one fracture set over another or the effect of an anisotropic state of stress on two fracture sets of different azimuths.

Fracture Aperture (e)

- **Aperture distributions from core, BHI & flow**

  – **Data Sets:**
  - Relaxed apertures in outcrop, core, thin section & CT scans
  - Partially stressed aperture from FMI determinations
  - Fully stressed aperture from flow coupled with fracture spacing and all flow comes from fractures (well test or laboratory)

  Increasing stress level and scale of measurement

Figure 8.3 An assessment of fracture aperture measurement sources as a function of stress state under which they are measured.

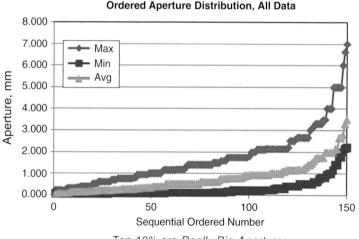

Ordered Aperture Distribution, All Data

Top 10% are Really Big Apertures

Figure 8.4 A typical distribution of mechanical-fracture aperture based on measurements of core from a carbonate section using a width comparator card from Ortega et al. (2006). Note that about 10% of the fractures were much wider than the rest of the data set. These would dominate fluid flow in the reservoir. In the core, the larger scaled fractures are missing from the data set as these are not adequately sampled by the core.

subsurface-related and not related to weathering at the earth's surface. If cores are unoriented we can gather data on the size variation of fracture apertures. In Figure 8.4 we see a core-based aperture data set from a limestone section capturing minimum, average, and maximum aperture from at least 150 natural fractures in a producing section.

Many companies do fracture analyses on Computed Tomography (CT) scans before they are taken out of the core barrels. As such, the core is somewhat more intact, and we can attempt to do work on natural open, and sometimes filled fracture, intensity in the relatively intact core. In terms of mechanical-fracture aperture, the measurement is still unstressed and there is an issue with resolution of the measurement and calibration. The resolution is not good enough to see the smaller fractures due to pixel size of the images and resolution of the CT images.

8.2   Partially Stressed

Partially stressed measures of natural fractures apertures can come from interpreted borehole image (BHI) logs. The primary tool used to gather this data is the Formation Micro Imager (FMI) log, but only if the operating company requests and pays extra for this data interpretation log tract. Because the measurement is obtained from a hole in the subsurface, the stress state operating on the fracture system is altered from that away from the wellbore, including hoop stresses. My understanding is that the technique measures the width of the resistivity signal along the sinusoidal trace of the fracture in the image and integrates the values along the trace. However, the width at any point of the fracture trace is dependent on the dip of the fracture planes and is especially in error at the top and base of the fracture trace where it interests the wellbore. Those measures are too large due to their oblique intersection of the wellbore and trace. They are also basically mechanical apertures. An example of image log-based fracture apertures is shown in Figures 8.5 and 8.6.

An alternative image log used to constrain natural-fracture apertures is the acoustic image log (ultrasonic borehole imager (UBI) and others). As discussed previously, these logs depict images of both travel time and amplitude of the inside of the borehole wall. The travel-time images are used to constrain natural-fracture mechanical aperture. However, the acoustic image log has half the resolution of the resistivity image logs, thus giving a less accurate measure than the resistivity image approach. In addition, the tool often requires relatively rough handling of the borehole condition to see the fracture planes adequately. The rougher hole will lead to greater error in the aperture calculation.

8.3   Fully Stressed

The final approach to measuring natural-fracture aperture for subsurface modeling comes from the fully stress condition of the reservoir. There are two types; one comes from wellbore flow test data or initial production decline in a double-decline production behavior constrained by subsurface measures on fracture intensity from core or perhaps image logs; while the

Natural Fracture Aperture Calculation from FMI

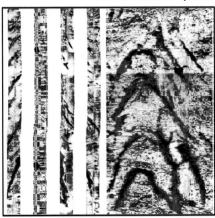
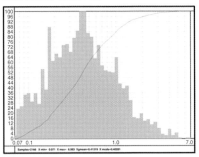

*Calculated FMI fracture
aperture distribution*

FMI Image   *UBI Image*

Figure 8.5   The traces of fractures in a borehole as shown on an FMI image (left), and on a UBI image (right). Both display the variable width (mechanical aperture) of the fractures along the fracture trace. Apertures are only measured from the FMI data and the value given for the fracture is the integrated width along the trace. The resultant aperture distribution diagram (right) displays a symmetric distribution which, according to the author, may not generally be the case. *Source:* from Ruehlick (2015), courtesy of B. Ruelick.

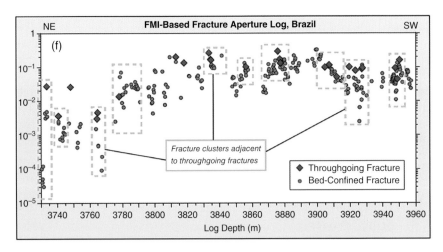

Figure 8.6   A fracture aperture data set for natural fractures in an offshore data set from Brazil, after Gross et al. (2009). It points out that two scales of fractures are measured (through going or cutting multiple beds, and bed-contained) and the larger through-going fractures have on average larger apertures showing a typical scaling relation. No discrimination by fracture azimuth is described. *Source:* courtesy of M. Gross.

second comes from laboratory measurements under simulated stress conditions for the reservoir.

In the flow test approach, we need a situation where we have a low-matrix permeability reservoir where over the length of the flow test, we can assume that all flow comes from the natural fractures present. In addition, we need some direct measure of natural-fracture intensity or spacing from the same zone of the flow test. The core or possibly image log fracture intensity can be applied. As such we can back out the hydraulic-fracture aperture necessary to give that flow rate, Nelson (1985). An example is given in Figure 8.7.

In the laboratory approach, we can look to Nelson (1985, 2001) for an approach. In this approach, we can approximate fracture aperture in the surface in laboratory-derived permeability tests. This is done by application of confining stress to companion samples (one with only matrix and one with matrix and a fracture) and measure the permeability under varying confining stress. We subtract the matrix-only permeability from the combined matrix and fracture permeability at every confining stress point and get the effect of the fracture. With an assumption of the fracture spacing (see Nelson 2001) we can back out an effective hydraulic fracture aperture as a function of depth or normal stress state; Figure 8.8a and b.

To complete the prediction of aperture in this approach, we need to predict which stress point on the curve is most appropriate for the depth of interest and possibly which points are applicable to different azimuths of fractures at that depth. What is the most appropriate point is that representing the normal stress component acting perpendicular to the fracture set of interest.

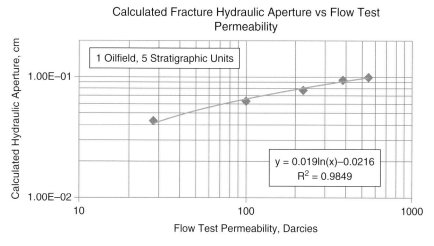

Figure 8.7 An average fracture aperture calculated from water flow test permeability from a shallow reservoir for five units with different average fracture spacing. These are calculated fully stressed apertures from the subsurface.

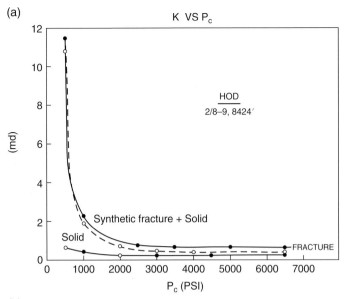

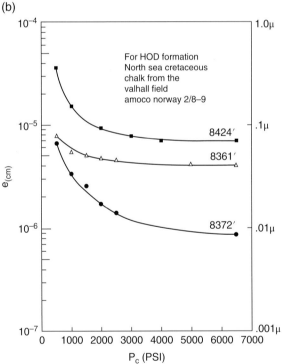

Figure 8.8 (a) The result of companion tests of permeability as a function of confining stress for a matrix-only plug and a matrix-plus fracture plug taken from core next to each other. This can be done for either natural fractures and synthetic fractures. The dashed line is the effect of the fracture at each pressure point. The rock is Cretaceous North Sea chalk. Source: figure is from Nelson (2001), p.75, courtesy of Gulf Publishing Co. (b) The effective hydraulic-fracture aperture for three such Cretaceous chalk tests as shown in (a). The assumptions needed to obtain the aperture are given in Nelson (2001). *Source:* figure is from Nelson (2001), p.77, courtesy of Gulf Publishing Co.

These aperture profile shapes and the order of magnitude of the apertures themselves are seemingly a function of lithology and grain size of the hosting rock with stiffer rocks displaying less severe fracture compressibility and finer grained rocks displaying smaller apertures.

8.4 How the Various Aperture Measures Go Together

As pointed out in the previous sections of this text, we can generate fracture apertures necessary to create static-fracture models for simulation from core/outcrop, image logs, backed out of flow tests/production data, and laboratory tests under confining pressure. These various apertures can be related as a function of stress level, Figure 8.9. This schematic Figure is from Nelson (2010d) and uses a lab-derived aperture vs stress curve as the central melding vehicle. In this diagram, the stress axis is related to the effective stress level in the subsurface due to depth but also the changes due to production and pore pressure reduction.

By knowing the stress level where the flow test/production analysis was performed and the wellbore hoop stresses where the image log calculations were performed, we can create this aperture curve by best-fitting the data. Alternatively, we can run the laboratory simulations to get the appropriated curve form.

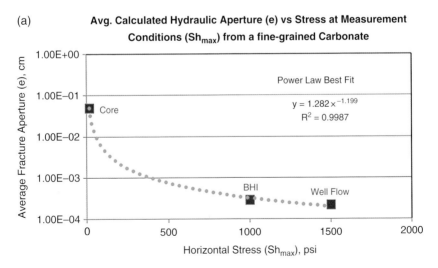

Figure 8.9 (a) The relationship between un-stressed, partially stressed and fully stressed fracture apertures (e) in terms of their relative position along a schematic laboratory derived aperture under stress curve, Nelson (2010d). (b) A schematic diagram depicting how the different data sets are combined to simulate burial and production.

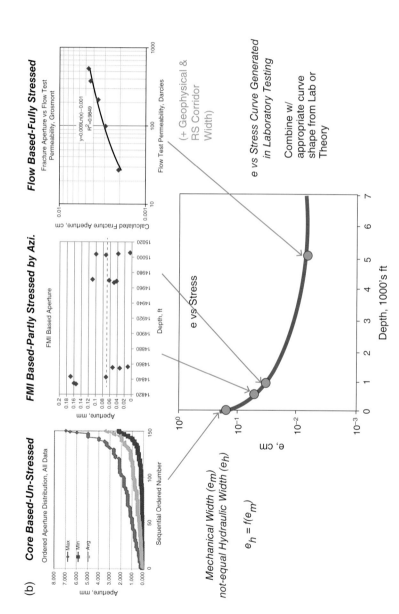

Figure 8.9 (Continued)

9

Creation of Natural Fracture Scaling Laws

When creating a Static Conceptual Fracture Model (SCFM) leading to simulation, it is best to model several scales of natural fractures, often three seems most appropriate. These could be through-going or bed-contained, mid-scale, and fracture corridors, etc., see Figure 9.1. Each would be modeled with their own orientation(s), spacing, height and length, and aperture. In general, the more numerous smaller fractures display the closest spacing and the smallest aperture, thus giving lower fracture porosity and permeability. Conversely, the largest features and corridors of multiple-fracture planes are much less numerous and have the largest spacing but have the largest cumulative aperture leading to very high fracture permeability. Intermediate scales of fractures fall relatively between those.

If we think about a car leaving a garage in the city, the car pulls onto an alley way that is very narrow and connected to other alleys. They are narrow, and the car moves slowly. At the end of the alley, it connects with a side road which is wider, and the traffic is now moving faster. Then the side road connects to a more major multi-lane wider road with much faster traffic. Finally, the car enters a major interstate highway that is wider, and traffic moves at its most rapid speed. This is a good model for fracture flow in a fractured reservoir. The scale of the fractures dictates the flow with the smallest of the interconnected system controlling the exit of oil and gas from the rock matrix while the largest set of the interconnected system controls the short-term very high-flow rate from a well that intersects it. Therefore, modeling must include the multiple scales of natural fracture development with some controlling recovery factor (smallest) and others controlling maximum flow rate (largest). This is the aim for our simulation predictions.

Static Conceptual Fracture Modeling: Preparing for Simulation and Development,
First Edition. R.A. Nelson.
© 2020 John Wiley & Sons Ltd. Published 2020 by John Wiley & Sons Ltd.

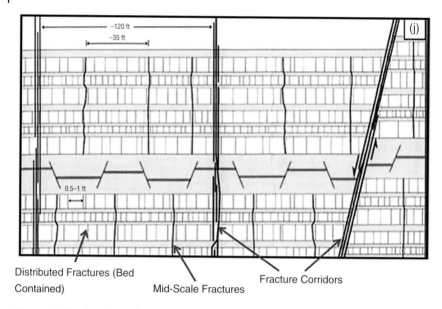

Distributed Fractures (Bed Contained)

Mid-Scale Fractures

Fracture Corridors

Figure 9.1 Depicted is a schematic geological cross section defined by mechanical stratigraphy variation and the three sizes of natural-fracture distribution found useful in modeling. *Source:* modified after Gross and Eyal (2007), courtesy of the Geological Society of America.

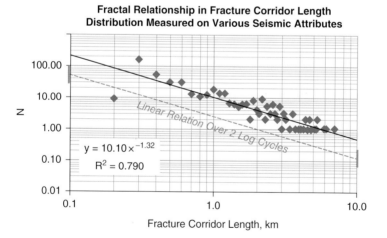

Figure 9.2 Shown is a typical fractal relation for lineation or fracture corridor length based on surface lineation and geophysical data. The slope of the linear best fit on this diagram is the fractal dimension.

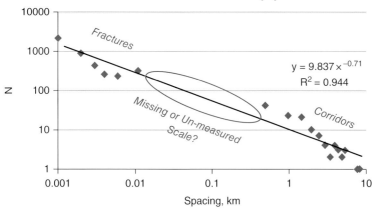

Figure 9.3 This diagram shows a fractal relation for fracture spacing over four log cycles. The larger are derived from geophysical derived fracture corridors while the smaller features are fractures derived from image-log interpretations in a horizontal well. From this relation we can predict statistics of intermediate scales of natural fracture sets for modeling purposes.

The properties of the natural-fracture sets or modeling parameters can show a variety of distributions; for example, log-normal, power law, or fractal. Many workers have shown that length or height, spacing, and aperture often display fractal distributions, Marrett (1997). This distribution is important for fracture modeling as we can often constrain the properties of fracture sets that we do not have the proper scale observations to measure.

A fractal distribution can be defined by plotting the number of features at a given size in a log–log plot, Figure 9.2. To be considered fractal, the plot should depict a linear relation over more than one log cycle with the slope of the linear portion being the fractal dimension.

In many instances, we can measure fracture property statistics in the wellbores (generally small scale) and from geophysical attributes (largest scale) but have little data on the statistics of the intermediate scale fracture features. An example of such a fractal relation that allows us to constrain the properties of an intermediate fracture set is given in Figure 9.3. With the relation and the fractal dimension we can generate fracture sets that we cannot measure but are consistent with the ends of the total distribution.

10

Gathering and Analyzing Mechanical Property Distribution Data

The distribution and orientation of natural fractures are controlled by the mechanical properties of the rocks in which they are developed and the stresses and strains that the rocks were subjected to. Therefore, to describe and potentially predict the fracture properties it is critical to be able to determine the mechanical properties of the rocks that make up our reservoir, as well as the layers that confine them or are interspersed with them.

In the oil and gas industry, mechanical rock properties are measured in two general ways, (i) static measurements based on finite strain in a rock mechanics laboratory, and (ii) dynamic measurements in the subsurface calculated from acoustic or sonic estimates derived with the assumption of infinitesimal strain and elastic behavior.

In the static measurements, we subject relevant rock samples from outcrop or core to various confining stress, pore pressure, temperature, strain rate, and interstitial fluid type to simulate subsurface stress conditions. Once under simulated burial conditions, the sample is generally loaded with a piston and the shortening and shape change of the sample is measured as external piston load is increased. Curves of the load (stress) and length or volume changes (strain) are created and various strength, stiffness, ductility and failure parameters are interpreted from the curves. Elastic or Young's Modulus (E) is defined in the tests as the linear slope of the stress–strain curve; while Poisson's Ration (γ) is the ratio of longitudinal to transverse length changes (length/bulging) during the static laboratory test.

In the dynamic measurements in the oilfield, we create elastic moduli by using sonic well log data. Of particular interest, are sonic logs that measure both primary (p-wave) and shear (s-wave) velocity. Using the assumption of elastic infinitesimal strain, *in-situ* elastic moduli (E & γ and others) can be calculated. This approach is very useful as we can determine the variation of these moduli on a fine scale in the rock units of interest relative to one another under *in-situ* conditions.

Static Conceptual Fracture Modeling: Preparing for Simulation and Development,
First Edition. R.A. Nelson.
© 2020 John Wiley & Sons Ltd. Published 2020 by John Wiley & Sons Ltd.

In most mechanical studies involving wells, we do a mix of both static testing and dynamic calculations from logs to calibrate the two approaches and see if there are any consistent differences. For most natural-fracture and hydraulic-fracture studies, the mechanical parameters most often utilized for determining fracture breakdown pressure and fracture containment predictions are Unconfined Compressive Strength (UCS), Young's Modulus (E), Rigidity Modulus (G), also known as Shear Modulus, and Poisson's Ratio (γ). Poisson's Ratio is expressed in percent (%), while UCS, E, & G are best expressed in Giga-Paschal (GPa), or alternatively psi $\times\ 10^6$.

10.1   Rock Modulus and How It Effects Deformation and Fracturing

Classically, mechanical property stratigraphic descriptions typically utilize a co-rendering of Young's Modulus (E) and Poisson's Ratio (γ) curves with depth in the wellbore. These moduli can be calculated in laboratory loading cells under finite strain, called Static Measurements, Figure 10.1a and b. Alternatively, we can calculate these moduli from shear sonic well logs, assuming elastic behavior and infinitesimal strain, called Static Measurements.

The two moduli measurements (Static and Dynamic) give slightly different values but are generally consistent in their variation. These calibrations are often shown in two ways. In one, a simple cross plot of both static and dynamic measures made on identical laboratory samples is created and a best-fit correlation curve is created. Examples of such cross plots for Static and Dynamic E & γ is given in Figure 10.2a and b.

An alternative method to display the Static and Dynamic measures is in strip log form, Figure 10.3. In this display, the basic shear velocity measures calculated for both approaches, are co-rendered with the Dynamic log-based data as the strip log base and the Static laboratory-based data plotted as individual depth-related data points along the strip log. Visually, we can see where the correlation is relatively good and bad.

In this presentation, we define the stronger more brittle layers by zones that display relatively high E and a corresponding relatively low γ. Alternatively, zones of relatively low E and corresponding relatively high γ are defined as less strong and somewhat more ductile, Figure 10.4. This is how mechanical stratigraphy is defined in a mechanical sense; brittle is high E and low γ and ductile is low E and high γ.

However, there are instances, where over some limited zones, the E and γ curves move not out of phase, but in unison, either up or down together. There must be a compositional and/or stratigraphic reason for this that is not yet evident at this time.

Figure 10.1 (a and b), the definition of Static (a) Young's Modulus (E), and (b) Poisson's Ratio (γ) derived from laboratory stress–strain tests under load.

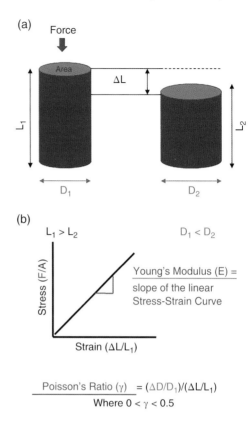

I prefer to use Rigidity Modulus or Shear Modulus (G) to define the mechanical stratigraphy of a section and define and predict subsequent fracture intensity (FI). It is also of use in defining rock facies and to refine stratigraphic tops.

G is calculated from the Young's Modulus (E) & Poisson's Ratio (γ) by $G = E/2(1 + \gamma)$ from either static or dynamic data, or from the density (ρ) and shear-wave velocity (v_s) from sonic logs by $G = \rho(v_s^2)$.

Rigidity Modulus (G), also known as Shear Modulus, is a measure of strength and brittleness of the rock package (higher Rigidity indicates a stiffer, harder, stronger, potentially more brittle material). We use it to define mechanical stratigraphy for prediction of deformation and brittle FI, Figure 10.5.

Rigidity responds to the make-up of the rock including its' mineral composition, porosity, grain size and fabric. There is an intimate relationship between the petrology of the rock package, its mechanical properties and subsequent natural FI for each generation of fracturing, provided deformation is primarily by a brittle fracture mechanism. The assumption needed to predict FI is that the variation in moduli we see today is the same variation that was operative when the rocks were fractured. This relation is less

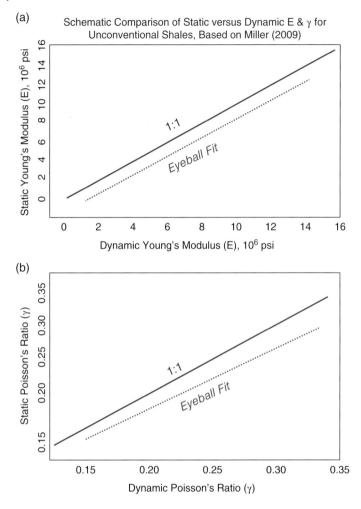

Figure 10.2(a and b) Static and Dynamic (a) Young's Modulus (E) correlation and (b) Poisson's Ratio (γ) correlation measurements on individual samples of a shale reservoir, data replotted from Miller (2014).

straightforward if strain partitioning takes place such that other competing deformation mechanisms are active, such as pressure solution and inter-crystalline twinning and translation.

Mechanical property distributions along a well (Mechanical Stratigraphy) also tells us something about the physical nature of the rock facies distribution and inferred changes in depositional environment, and how they change well to well. It seems to define internal rock character better than typical single parameter log-based approaches such as gamma ray logs. The approach can

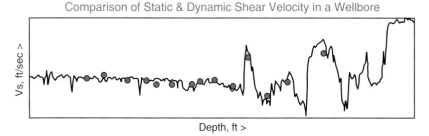

Figure 10.3 Strip-log correlation between Static (line) and Dynamic (blue data points) for Shear Wave Velocity as a function of well depth. A good correlation lends credence to our log-based data base.

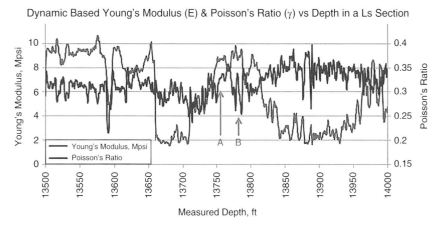

Figure 10.4 A typical plot for Dynamic Young's Modulus E and Poisson's Ratio (γ) for a well in a Jurassic Limestone package. In general, the interpreted stronger, stiffer, more brittle units are where the two curves are distinctly out-of-phase (B as example). However, especially in carbonates, we often see small zones where the two curves are in-phase (A as example). At present the in-phase behavior is ill-defined.

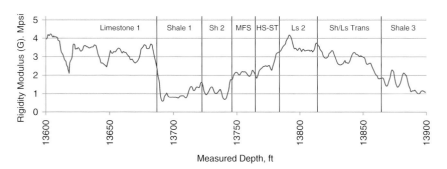

Figure 10.5 Rigidity Modulus (G) for the same well data as in Figure 10.4. G Modulus is calculated from the E & γ curves using $G = E/2(1 + γ)$. This is, perhaps, a better representation of the mechanical property variation in the reservoir section.

more accurately define edges of facies than in visual description and can show similarities or differences in facies and/or environment of deposition between wells. If available, have the Rigidity curve on hand when doing sedimentological/ stratigraphic description of core as an aid or guide.

10.2 Rigidity Modulus Distributions

The concept of Mechanical Stratigraphy inherently derives from a consistent variation in both mechanical properties and deformation mechanism and, therefore, natural FI through the rock package. Along wellbore mechanical property variations of E & γ or G logs are used to define mode of deformation in vertical wells, or at a finer stratigraphic scale, in horizontal wells. The well-bore measured G values can be handled and plotted like FI to show variation between wells in a field, between field sectors, and between fields, as shown in previous Figures 4.17–4.19.

Indeed, mechanical property sampling via wireline logs or cores in a vertical well emphasizes variation formation to formation for an entire rock section of interest. That in a horizontal wellbore, on the other hand, tends to emphasize finer-scale variation within a single reservoir, usually facies or sub-unit variation as the wellbore undulates up and down along its track.

10.2.1 Vertical Distribution in Wells

Mechanical property logs are usually run in a vertical well over a relatively long zone of the rock section to sample not only the target reservoir, but stratigraphic units above and below as well. The purpose is to investigate property variation between formations or stratigraphic units. Ideally, this sampling would include both reservoir and seal, as this can help constrain prediction of hydraulic fracture containment during well completions. An example of mechanical property distribution in a vertical well plotted with a core-based FIC is shown in Figure 10.6. Note that there is a good correspondence between the peaks of the two curves as well as the shapes of those peaks. This North American data set allowed for a numerical relationship between G and FI to be created and used to "predict" FI in un-cored wells from mechanical properties logs. This is important as it is difficult and expensive to obtain core in many wells during exploration and development.

The core FI and Log G correlations represented in Figure 10.6 were added to 11 other cored wells to create a best-fit equation for use in predicting *in-situ* FI directly from Shear Sonic log data G values, Figure 10.7.

A common practice for drilling a horizontal well in order to maximize natural fracture interception rate is to first drill a vertical pilot hole from which the horizontal segment will be drilled. The vertical portion is often drilled deeper than the eventual target for the horizontal segment and the vertical well is

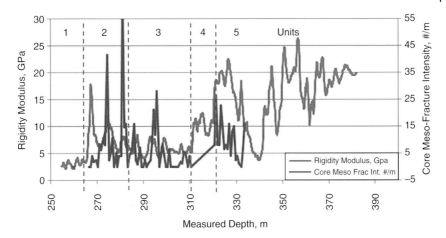

Figure 10.6 This is a co-rendered plot in strip-log format showing variation in both Rigidity Modulus (G) and core-based natural fracture intensity (FI) down a vertical wellbore. Stratigraphic units are listed as numbers for further reference. Notice the similarity in the peaks of the two curves. From data like this, the two data sets can be calibrated with one another and an algorithm for Fracture Intensity calculation from G created. However, this algorithm will only be appropriate for this section in this area.

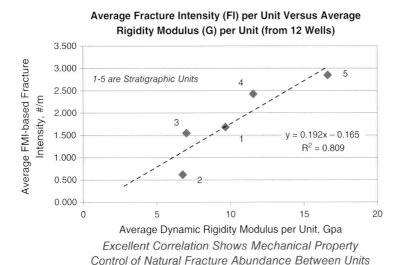

Figure 10.7 This cross plot represents the average image log Fracture Intensity (FI) vs the average Rigidity Modulus (G) for five separate stratigraphic units in a reservoir section as in the previous figure, averaged from 12 wells. These are the same units shown in Figure 10.6 and the cross plot is validation for the algorithm created for prediction.

quickly investigated for mechanical property variation and evidence (from logs or cores) of *in-situ* natural fracture orientation the magnitude and the orientation of *in-situ* subsurface stress components. With this data from the vertical pilot hole, the depth and azimuth of the horizontal wellbore is determined.

10.2.2 Horizontal Distribution in Wells

Theoretically, a horizontal well segment is either drilled to be geometrically horizontal or to exactly follow one optimum stratigraphic horizon. In practice, however, drilling a horizontal well is not exact, primarily because it is difficult to determine exactly where the bit is at any time during drilling. The result is a wellbore that rises and falls along the reservoir path. The wellbore is said to "porpoise" like the aquatic mammal does while swimming alongside a boat. The result is a mechanical property variation usually within the broader reservoir, but sampling varying facies or subunits around the target horizon, Figure 10.8.

The high and low portions of the porpoise-like wellbore have both good and bad effects. On the positive side, a long horizontal well segment may cross the reservoir top six or eight times along its length. Interpreted correctly from the

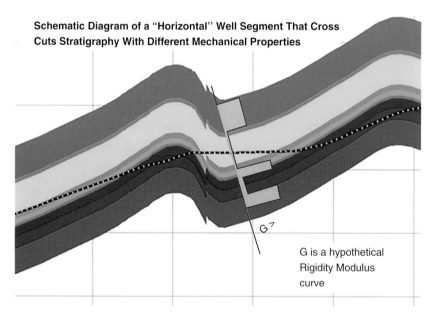

Figure 10.8 A schematic diagram depicting a typical well path for a lateral well. The path tends to rise and fall stratigraphically as it is hard to keep the well exactly straight, or as shown here local structure can cause the well to sample various units along its' path. The result is a log suite that is difficult to interpret and potentially significant differences in measured mechanical properties and natural-fracture intensity. Base is a photo adapted from the Stoner Engineering website (https://makinhole.com/SES_Geosteering.htm).

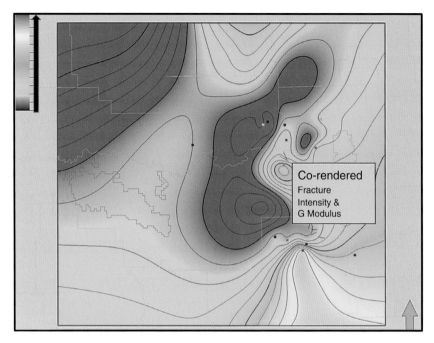

Figure 10.9 A co-rendered map of Rigidity Modulus (G) in color contours with Core-based Natural Fracture Intensity in black-line contours for a productive area in the western US. The shapes of the two maps are almost identical indicating a near 1:1 correlation between G Modulus and Fracture Intensity in map form.

logs, this gives many more data points from which to map our reservoir surface. One negative feature of the rise and fall of the wellbore trajectory is on hydrocarbon flow during production. These relative lows can collect water and act as "sumps" obstructing hydrocarbon flow along the wellbore. A second negative is that without knowing how the wellbore is sampling the rock section, we may misinterpret the reservoirs petrophysical properties from our well logs.

These within-reservoir variations in mechanical properties can help guide selection of fracture initiation points and aid in defining natural FI predictions to help predict and interpret variations in the vertical and horizontal growth of the hydraulic fractures.

10.2.3   Map Distributions by Unit and Sub-units

By creating average Rigidity or G values for each important stratigraphic unit of interest from many wells, maps can be created depicting variations in mechanical properties laterally along horizons, as well as variations at any selected position (potential drilling locations) between horizons. An example of such a well-constrained map of G co-rendered with core-based average natural FI is shown in Figure 10.9. Such maps are very useful in infill drilling of a productive trend involving natural fractures.

11

Locating Fracture Corridors

As seen in Figure 9.1, "Fracture Corridors" are swarms of fractures usually 10s to 100s of meters wide, often spaced at regular intervals with internal steeply dipping fractures. In addition, some corridors are related to faults of all types. As pointed out by Richard et al. (2017), tall fracture swarms are often the best-connected fracture systems in a reservoir and, therefore, dominate fluid flow if encountered by our wells. Quantification of these swarms is very important in constructing a useful Static Conceptual Fracture Model (SCFM).

One of the most direct methods to determine the width of natural-fracture corridors is by observations in horizontal wells. This can be accomplished with the observation of horizontal core taken over relatively small sections of the horizontal well, or with acquired borehole image logs (BHI) and interpretations over longer portions of the well.

Figure 11.1 shows how we can determine corridor widths in three horizontal wells in the Middle East. We see in this figure three well paths, the position of two seismically mapped faults, wireline logs, and natural fracture intensity curve (FIC) along the wellbores. Using 3D integration like this allow us to determine the width of corridors associated with the faults.

In another example, BHI interpretations in horizontal wells can be turned into a FIC and displayed along the length of the wellbore, Figure 11.2. In this figure, there are many thousand data samples making it difficult to see "the forest for the trees." Therefore, the data is smoothed using a boxcar moving average technique. Shown in this figure are two versions of smoothing; 1 smoothed using an 11-sample window and another using a 201-sample window. In the 201-sample smoothed curve we can see six separate natural-fracture swarms that are interpreted to be related to individual faults. These faults are confirmed using the BHI-derived bedding dip interpretations in the well.

These data sets are best generated from various geophysical data, interpretations in deviated and/or horizontal core and BHI. In addition, these distributions can be constrained by utilizing published outcrop-based models.

Static Conceptual Fracture Modeling: Preparing for Simulation and Development,
First Edition. R.A. Nelson.
© 2020 John Wiley & Sons Ltd. Published 2020 by John Wiley & Sons Ltd.

**Fault Zone Injector/Producer Short-Circuit with
Diagenetic Effects**

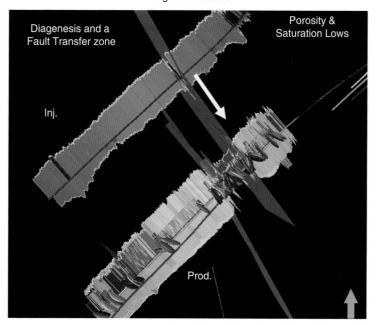

Figure 11.1 A unique example of data integration in 3D from a Petrel image showing three wellbores (one injector and two producers) with porosity and saturation logs, and a Fracture Intensity Curve (FIC). There is a breakthrough or "short circuit" between the injector and producers. The gray planes are seismically mapped faults. We see a displacement transfer zone in the producers with high-intensity fractures at the faults and in the transfer zone between them. These have been diagenetically altered by fluids moving up the faults reducing the porosity and saturation of the fault-related fractures by mineralization in that zone.

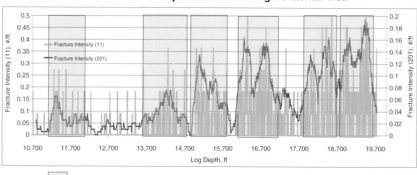

Figure 11.2 An FIC from a long 9000' horizontal well segment in the Williston Basin showing six interpreted strike-slip faults. The curve is smoothed using a 201-sample window and a sample length of 1' using BHI fracture interpretations. Corroboration comes from horizontal core in the first of the interpreted faults. *Source:* from Buckner et al. (2013) courtesy of Marathon Oil.

Further evidence of fracture corridors containing a high intensity of connected fractures can come from the position of fluid-flow entry into a wellbore. This is done with the use of a Production Logging Tool (PLT). This log shows fluid entry into the wellbore while the well is flowing. There are four kinds of PLT; rate, temperature, acoustic, and borehole television (BHTV). The PLT rate approach is called a "spinner survey" and basically uses an anemometer measuring fluid-entry rate. It is mostly used for oil flow. The curve depicts an ever-increasing rate from the bottom of the logged interval to the top of the interval, Figure 11.3. Lengths of the log where there is a rapid slope increase in rate are where the

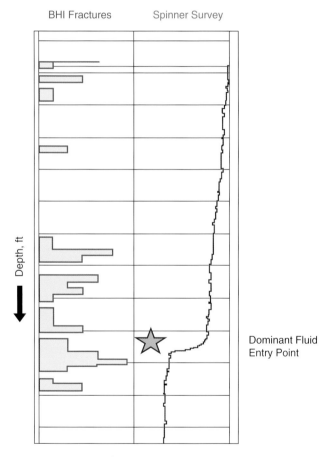

Figure 11.3 Two logs from a vertical well in South America; Fracture Intensity Curve (FIC) from UBI interpretations, and a Production Spinner Log. There are numerous fracture swarms seen in the wellbore, but only one shows significant fluid entry into the wellbore at the depth marked with the star. This indicates that only this one zone contains properly connected fractures or that the other mapped fracture zones have been damaged by invasion of drilling mud. A mud-loss curve would aid in determining which is the case.

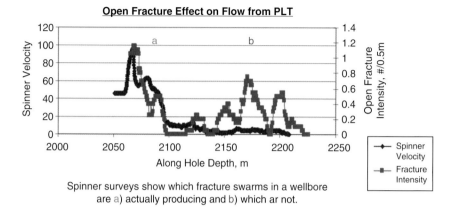

Figure 11.4 A fracture intensity curve (FIC) and a spinner velocity plot for a flowing well in North Africa.

fluids from the reservoir are preferentially entering the wellbore. Notice in this figure that not all fracture zones flow oil. Indeed, in this field only one or two of the numerous fracture zones identified by ultrasonic borehole imager (UBI) interpretations in the wells flow hydrocarbons, and those zones produce 70–80% of the total production. This is the case in all fractured reservoirs. High numbers of fractures at the wellbore surface does not mean that those fractures are connected to others away from the wellbore. Those that are connected and flow, are likely to be associated with large-scale fracture corridors.

Another example of a PLT spinner survey response in a vertical well with natural fractures is shown in Figure 11.4. Once again, several high-intensity fractured zones in the lower portion of the survey show no flow while the high-intensity zones at the top of the survey do.

In terms of the other PLT types, the temperature PLT, measures fluid temperature along the producing interval and is used mostly in natural gas wells. While the fluid coming from the formation should be hot, the gas entering the wellbore expands and cools due to the Bernoulli Effect. The result is relatively cool zones of gas production. The acoustic PLT involves a microphone drawn up the wellbore while it is producing to "hear" fluid entry. Once again this is predominantly a tool used in gas wells. The BHTV approach to PLT gives a television picture of the wellbore as production is taking place. It relies on being able to see particulate material moving within the flowing oil or gas. The downside is that the borehole must be filled with clear fluid while the log is being run.

The wellbore position of faults and their associated fractures can be inferred from structural data in selected wireline logs. This can be done by older Dipmeter Logs displaying empirical dip patterns of faults of various type, or now days by borehole image logs, like FMI or UBI and others. For natural- fracture modeling, we generally work with just the fracture position and attitude interpretations.

However, there is also great structural information in the bedding dip interpretations. Indeed, profiles of bedding dip azimuth and dip magnitude display discontinuities in the bed geometry that can be interpreted as faults. These discontinuities often show up as "no interpretation" zones as bedding becomes chaotic in the fault zone and inherent fractures are hidden by the disruption or are deformed. Therefore, the width of the no interpretation zone can be considered the maximum width possible for the fracture corridor associated with the fault.

Figure 11.5a shows an FMI-based example of dip magnitude data depicting a fault and the potential width of the related fracture or damage zone. As part of

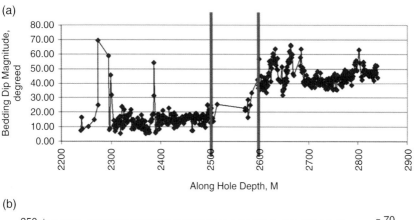

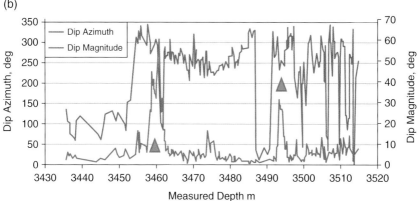

Presence of 2 Potential Faults? (3460 & 3495 m), Corresponds to Clusters of Micro-Faults Interpreted on the Image Log

Figure 11.5(a and b) BHI-based interpretation of bedding strike and dip can be used to define local structure encountered by a wellbore. In (a) we see a large disruption in the magnitude of dip of 30° over 100′ of wellbore. Few bedding picks are available over this interval indicating disruption of bedding in a fault zone. There should be natural fractures associated with this fault. In (b) bedding dip magnitude and dip azimuth are plotted along the wellbore path and depict the general position of two faults, highlighted by green triangles. Fault drag is also evident. Once again, there should be natural fractures associated with these faults.

Auto-picked seismic discontinuities overlain on a Curvature-on-Coherency display. Both are Seismic Discontinuity Attributes and Show Good Correspondence.

The concept is that we are finding the fracture corridors and the parallel dark lines as showing the edges of the high intensity fracture zone. These widths can be measured at numerous points along the trace and a Fault Damage Width Distribution for each azimuth class constructed.

Figure 11.6 Geophysical attributes that can be used to estimate fracture corridor width distributions by measuring the distance between parallel red lines (highly curved edges) at multiple positions along the features as shown by the green arrows. *Source:* adapted from, Wagner et al. (2010).

the fault inference is the large dip difference on either side of the no-interpretation zone (40° dip difference). Figure 11.5 b shows similar fault interpretations from both bedding dip azimuth and dip magnitude. Natural fracture corridors will be associated with each of these faults and incorporated into the SCFM.

The placement of faults, including their strike and dip and sense of motion in the model, is best done directly from 3D seismic surveys with the constraint of wireline logs. However, the width of the damage zone or elevated fracture intensity zone surrounding the fault from seismic data is much more difficult and requires manipulation of the seismic data in the form of seismic attributes. An example of seismic depth curvature with automated discontinuity mapping is shown in Figure 11.6. This type of data allows us to map a view of the width and width distribution of fault zones for input to SCFM creation.

12

Rock Anisotropy and its Importance in Determining Dominant-Fracture Orientation and Relative Intensity

Naturally fractured reservoirs usually display horizontal anisotropy in *in-situ* permeability and production rates and volumes. This azimuthal anisotropy is caused by two factors by themselves, or a combination of the two. First, the natural-fracture distribution may be dominated by vertical open fractures of one dominant azimuth, with small dispersion in strike. This could be a Systematic Regional Fracture system. As open fractures are very high in permeability parallel to the fracture plane, and generally much higher in permeability than the matrix, this becomes the dominant flow direction in the reservoir. The result is often highly elliptical drainage areas in fractured reservoirs, Nelson (1985). The second factor that can create the anisotropy in flow and drainage is the *in-situ* state of stress. If we had a fold-related natural-fracture system with six orientations of potential fracture planes all perpendicular to folded bedding, our flow could almost show isotropic flow and drainage. However, the two horizontal principal stress components in the reservoir are generally different in magnitude and will compress some fracture azimuths more than others. As the permeability of a fracture is a cubic function of its aperture or hydraulic width, those natural fractures perpendicular to the maximum horizontal stress component in a relaxed basin will be relatively more closed or of smaller aperture and, therefore, lower in fracture permeability; while those fractures perpendicular to the minimum horizontal stress component will be relatively more open and of higher fracture permeability. As a result, the *in-situ* stress state can turn a complex natural fracture distribution into one with strong geophysical and permeability anisotropy, Figure 12.1.

Much work has been done over the last 20–30 years trying to measure horizontal anisotropy in fractured reservoirs from geophysical data, both from 3D seismic surface-based surveys and in wellbores using wireline logs. The concept applied in interpretation is that seismic waves will be retarded or slowed in their travel time when crossing an open fluid-filled fracture; while not noticeably retarded or slowed in the direction parallel to the open fracture. In the interpretation of the data, mineral-filled fractures are considered to have

Static Conceptual Fracture Modeling: Preparing for Simulation and Development,
First Edition. R.A. Nelson.
© 2020 John Wiley & Sons Ltd. Published 2020 by John Wiley & Sons Ltd.

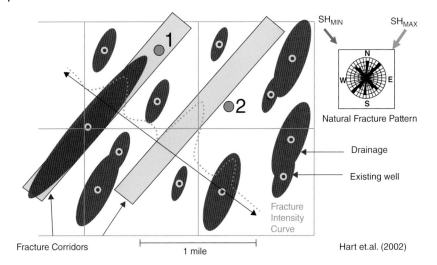

Figure 12.1 A schematic depicting elliptical well drainage areas due to preferential fracture permeability direction due to *in-situ* horizontal stress components. Only the natural fractures perpendicular to the minimum principal stress dominate the fluid flow. In addition, it shows the effect of Fracture Corridors or fracture clusters on interconnected permeability and well drainage area. The biggest production will occur in Fracture Corridors that are intersected by a wellbore. *Source:* figure is modified from Hart et al. (2002), courtesy of Geoscience World.

no effect on the seismic velocity. However, we can envision that mineral fills that are dramatically different mechanically from the host rock might cause reflection of the seismic waves, thus, possibly looking like open fractures. To my knowledge, no research has been reported on this potential reflectivity.

To some degree, resultant open-fracture direction determinations from seismic data is often moot. If we understand the natural-fracture distribution and morphology from other data, such as core and outcrops and structural mapping we may already anticipate the fluid-flow anisotropy. However, without such fracture data, flow anisotropy from seismic data can be very beneficial during development drilling and especially during secondary recovery planning.

I have had experience with very good wellbore sonic logs, including shear sonic logs, to characterize mechanical stratigraphy and natural-fracture distribution. When supported with independent natural-fracture characterization and production analyses, these logs can indeed be used successfully to estimate the azimuth of near wellbore maximum horizontal permeability. An example of such an approach in a known fractured reservoir is given in Figure 12.2. This figure plots a shear sonic anisotropy log which depicts the strength of the horizontal sonic anisotropy, along with a mechanically predicted fracture intensity curve (locally calibrated G and FI relation) and a Stonley Wave curve that is

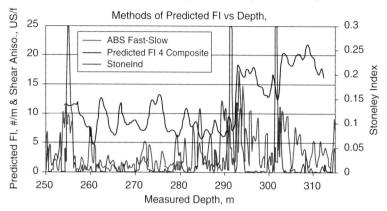

Predicted FI, Shear Anisotropy & Stoneley Index vs Depth

(Shear Anisotropy & Stoneley Index Highest at Unit Boundaries (stress concentration due to elastic miss-match?)

Figure 12.2 This diagram plots three data curves along a vertical wellbore in North America. They are a sonic anisotropy curve (difference between the fast and slow directions), a predicted Fracture Intensity Curve (predicted from a core FI and G calibration), and a Stonley Wave curve. The Stoneley Wave curve is intended to measure permeability along fractures away from the wellbore surface. Where all three curves are relatively high, higher-intensity fracture zones are predicted from this data. The three all peak at unit boundaries, perhaps because elastic mismatch at the boundaries causes stress concentrations and, therefore, more fractures.

purported to highlight zones of fracture permeability outside of the wellbore, particularly fracture permeability, Ward et al. (2014). This figure shows the general aspect of fractured reservoirs mentioned previously, that only limited zones in the wellbore fracture distribution produce, Rawnsley (2007). These logs correspond well, but it does not always work so well, and none should be used by themselves alone. In general, no single wireline log approach should solely be used to determine natural fracture fluid flow zones. Parameter compilations seem to work best in fractured reservoirs where we look for the preponderance of response within a group of predictors.

Scaling up in investigation to 3D seismic surveys gives a broader view of the anisotropy and samples a much larger volume of reservoir than the near well-bore vicinity. As a result, 3D survey sonic anisotropy analyses are a paraphrase or high-level view of the actual natural fracture system, ignoring the individual complexity at any given point in the reservoir, i.e. what is the stress state, number of fracture sets and their abundance, and fracture morphology distribution. The analyses will give only one resultant open-fracture direction out of the complex reality. It is difficult to determine how strong a control this direction will have in determining overall flow anisotropy.

13

Determine the *In-situ* Stress Directions and Magnitudes and their Variation

The subsurface *in-situ* stress state is an important aspect in understanding and quantifying fractured reservoirs as it directly impacts the behavior of a reservoir and its fractures prior to, and during, production. Stress conditions in the reservoir dictate the mechanical behavior of reservoir rocks during deformation. It does so by controlling the relative stiffness and ductility of the rocks during deformation and fracturing. Perhaps a greater effect comes during fluid production, hydrocarbon or water, where reservoir compressibility can alter both matrix and natural fracture fluid flow properties. Indeed, the compressibility of natural fractures is significantly higher than that of the matrix (Nelson 2001) and so the high-flush fracture permeability will decrease more rapidly than that of the matrix, Figure 13.1. This can destroy a high-rate fractured reservoir production rate in a relatively short duration of production. If depletion is too rapid, the fractures can narrow or close due to pore pressure reduction. In such reservoirs, rate control may be of paramount importance.

In terms of natural fractures within a reservoir, fracture porosity (ϕ_f) and fracture permeability (k_f) are both a function of fracture width or aperture (e) of the fractures and the abundance or spacing of those fractures (D), Nelson (1985). ϕ_f is a direct function of e, while k_f is a cubic function of e, see Chapter 8 on Fracture Aperture. Therefore, the orientation and relative magnitudes of the principle components of the *in-situ* state of stress will have a much greater effect on k_f than on ϕ_f. Thus, the *in-situ* stress state will act to close the fractures in the reservoir or reduce their aperture in some directions more than others. This results in great permeability anisotropy and elliptical drainage in fractured reservoirs.

For reservoir simulation modeling and prediction of flow rates, well paths, and recovery factors, it is imperative that we first know what the orientation of the *in-situ* principle stress components are in our reservoir and what the magnitudes of those components are, Figure 13.2. The processes involved in these determinations are very well described in Zoback (2007).

The remainder of this section will show examples of these types of data and how they are displayed for modeling purposes.

Static Conceptual Fracture Modeling: Preparing for Simulation and Development,
First Edition. R.A. Nelson.
© 2020 John Wiley & Sons Ltd. Published 2020 by John Wiley & Sons Ltd.

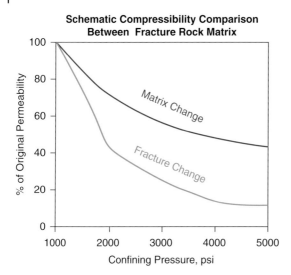

Figure 13.1 Schematic diagram showing the relative difference in compressibility between a sandstone matrix rock and a natural fracture within that rock. This relative difference is based on laboratory data and is typical for all fractured reservoirs with open, not propped fractures.

13.1 SH$_{max}$ Directions and Mapping

Assuming a "relaxed basin" with burial stresses only and no significant tectonic stress overprint, the maximum principle component of the *in-situ* stress state will be vertical. If there is no mechanical anisotropy, the two horizontal principle stress components should be less than the vertical and equal to each other. However, due to differences in mechanical properties azimuthally and perhaps mechanical fabric of the reservoir, there is usually a difference between the two horizontal principle stress components. The greater the difference between the two, the greater the resultant permeability anisotropy due to natural fracture systems is likely to be.

Several techniques generally used to determine stress component orientations and magnitudes are shown in Figure 13.3.

Most frequently, drilling-induced fractures and borehole breakouts from image logs and core are the techniques of choice for determining the orientation of the principle stress components. As shown in Figure 13.4a and b, drilling-induced fractures seen in core and BHI parallel Sv and SH$_{max}$ and are perpendicular to Sh$_{min}$. Borehole breakouts on BHL, on the other hand, parallel Sv and Sh$_{min}$ and are perpendicular to SH$_{max}$. If image log interpretations are done correctly, and they are sometimes not, the azimuth of breakouts and drilling-induced fractures should be perpendicular to one another. If not, there must be an interpretation error in one or both features. This is a good quick quality control check on the interpreted data.

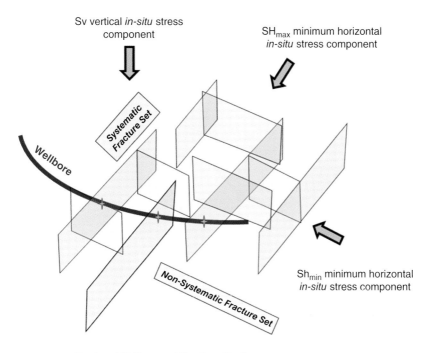

Sv vertical *in-situ* stress component

SH$_{max}$ minimum horizontal *in-situ* stress component

Shmin minimum horizontal *in-situ* stress component

Regional Orthogonal Fracture System

Figure 13.2 A schematic of a horizontal well segment penetrating a system of regional fractures with the systematic set in blue and the non-systematic set in green. Also shown are the *in-situ* stress components in relative magnitudes. Those vertical fracture planes that are perpendicular to the lowest horizontal stress component will be the most open, while those perpendicular to the highest horizontal stress component will be the most closed or lower in fracture permeability. *Source:* figure base is after Weng et al. (2011), courtesy of Society of Petroleum Engineers.

Methods to Determine *In situ* Stress Directions

- *Published stress maps (World Stress Map)*
- *BIL drilling induced fractures (parallel to Sh$_{max}$)*
- *BIL borehole breakouts (parallel to Sh$_{min}$)*
- *Numerical modeling using discontinuities*
- Strain relief measurements on core
- Door stopper technique
- Azimuth control of fracture aperture from BHI or sonic logs

Figure 13.3 Listed are the ways we can approximate the orientation and possibly the magnitudes of the *in-situ* principle stress components.

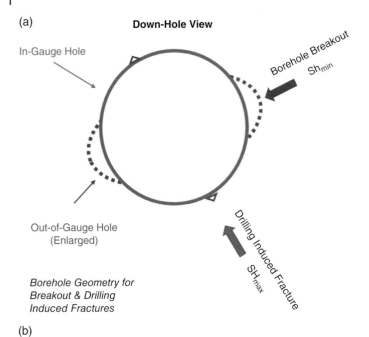

(a)   **Down-Hole View**

In-Gauge Hole

Borehole Breakout
Sh$_{min}$

Out-of-Gauge Hole
(Enlarged)

Borehole Geometry for
Breakout & Drilling
Induced Fractures

Drilling Induced Fracture
SH$_{max}$

(b)

Horizontal Stress Component Directions from FMI Interpretation in Appraisal 5 Wells
Sh$_{max}$ = c.50° Azi.

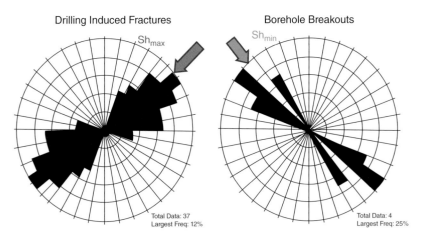

Drilling Induced Fractures   Borehole Breakouts

Sh$_{max}$ Sh$_{min}$

Total Data: 37   Total Data: 4
Largest Freq: 12%   Largest Freq: 25%

Figure 13.4(a and b) Differences between borehole breakouts, drilling-induced fractures, and the *in-situ* stresses for a single well in a) the borehole geometry, and (b) Rose Diagrams of BHI interpretations for drilling-induced fractures and borehole breakouts. The two data sets should be 90° apart in azimuth or one or both data sets are questionable. The result would be reinterpretation of the data, or a high risk of stress component azimuth.

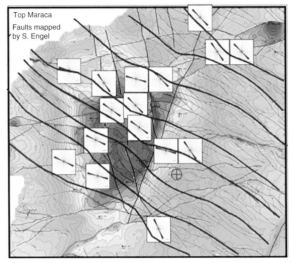

Poppelreiter et.al. (2005)

Figure 13.5 A stress trajectory map for Urdaneta West Field in Lake Maracaibo, Venezuela. What is plotted are individual Rose Diagrams of the azimuth of the Maximum Horizontal Stress Component as interpreted from ultrasound borehole imager (UBI) image logs. The thick blue lines represent trajectory lines drawn by hand for the SH_{max} data. This direction is the current maximum stress component in a strike-slip state of stress. *Source:* the figure is after Poppelreiter et al. (2005). Reproduction Courtesy of AAPG.

Having made such interpretations from formation micro imager (FMI) data, maps can be created to show the in-field variation in *in-situ* maximum horizontal stress component direction including stress trajectories, Figure 13.5.

As pointed out previously, *in-situ* stress state data can be generated from a reservoir study in a single area or come from available data collections, such as the World Stress Map or other published regional compilations. Such a local compilation is given in Figure 13.6.

As shown in this figure, multiple types of data sources can be integrated effectively to create a stress orientation map useful as a starter for fracture modeling in a prospective area. Sources of these compilations typically include earthquake data, BHI interpretations, structural mapping, outcrop measurements, and core measurements.

13.2 SH_{max} Directions with Depth

In areas experiencing a current tectonic stress state or a relaxed one with a residual or "locked-in" stress state that has not been totally relaxed, we do see variations in SH_{max} directions and magnitudes with depth or stratigraphy.

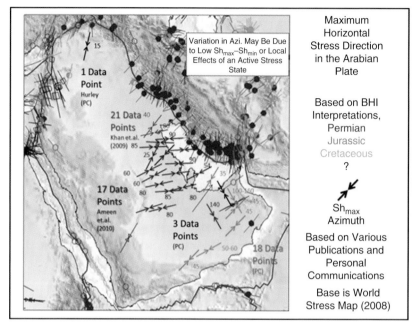

Figure 13.6 This map is an example of a stress orientation compilation by this author with orientations plotted on a portion of the World Stress Map (2008) www.world-stress-map. org/download. It is the type of compilation map useful to assist with stress-state determinations in any area for fracture modeling. The area of the figure centers on the Arabian Peninsula. Plotted is the azimuth of the current Maximum Horizontal Stress Component (SH$_{max}$) from several sources; World Stress Map (2008), Khan et al. (2009), Ameen et al. (2010), and several Personal Communications. *Source:* this Figure is after Nelson (2010a).

An example is shown in Figure 13.5 for the well; location marked with a cross. The two Rose Diagrams of SH$_{max}$ azimuth from BHI interpretations at that well are taken from two sides of an active E-W strike-slip fault. They show a shift of about 40° azi. The fault initiated in the Jurassic and has seen two periods of reactivation with changes in motion direction since then. The result is a rotation of the *in-situ* stress field caused by the fault. Such rotations must be taken into account when modeling fracture orientations and fracture permeability in the Static Conceptual Fracture Model (SCFM) process.

This can be investigated further by creating plots of SH$_{max}$ direction from BHI interpretations with depth and stratigraphy, Figure 13.7a and b. In Figure 13.7a, a relatively small variation in azimuth is seen and this +/− variation can be used as a measure of precision or accuracy in log interpretation, thus quantifying stress direction risk. However, in Figure 13.7b, we see a variation in azimuth that is abrupt and of 20° azi. This change is interpreted to be a fault and when

Examples of Sh$_{max}$ Azimuth with Depth From Borehole
Breakout Interpretations

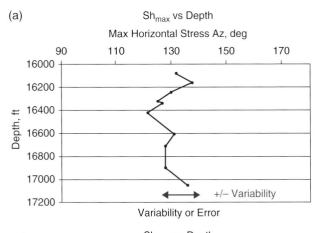

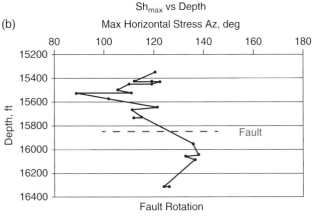

20° shift in azimuth at about 15,000

Figure 13.7 When interpreting the azimuth of the Maximum Horizontal Stress Component (SH$_{max}$) from image log borehole breakouts or drilling-induced fractures, it is recommended that the interpreted azimuths be plotted with depth in the vertical well. The reasons are twofold, (a) doing so will give an estimate of +/− error in the interpretation, and (b) showing abrupt changes in azimuth can help interpret the position of faults in the wellbore and a "critically stressed" fault.

seen is usually accompanied with BHI-interpreted bedding strike and dip changes. These SH$_{max}$ azimuth variations at faults are very important in terms of modeling natural-fracture permeability anisotropy direction near faults.

In conclusion, we can see that the relative azimuth and magnitude of the principal *in-situ* stress components and that of the dominant open-fracture

**Resolved Stress on Mapped Faults
Converted to Permeability**

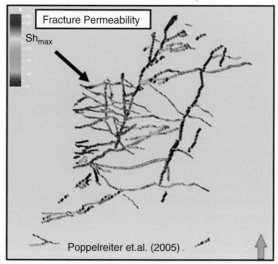

Figure 13.8 A map of calculated fracture permeability associated with reservoir faults in the Urdaneta West Field, Lake Maracaibo, Venezuela. It represents a manipulation of the current *in-situ* stress field and fault azimuth for every pixel along the trace of the faults. Using a published algorithm from another productive area, this is converted to fracture permeability. *Source:* the figure is from Poppelreiter et al. (2005), Courtesy of American Association of Petroleum Geologists.

azimuth controls the distribution of fracture aperture azimuthally and, therefore, the resultant maximum permeability direction for creation of an SCFM. An example of such modeling is shown in Figure 13.8. In this figure, data of the interpreted stress variation across a portion of the Maracaibo Basin is compared with the azimuth of mapped faults in the reservoir on a grid-square by grid-square basis, and the resultant *in-situ* stress component acting perpendicular to the fault trace resolved. Using a published algorithm from the North Sea, this was converted into fracture permeability for creating the SCFM.

14

Production Calibration

An important aspect in the static fracture modeling as well as in the simulation history matching is the correlation to production data. That data, except for early well testing, is only generally available later in the development phase of the project after enough production has been metered and analyzed. However, its relation to the static model is, after all, the point of the simulation. In my experience, there are several production parameters that have proven useful to calibrate to natural Fracture Intensity; Figure 14.1.

The correlation or calibration between the production data and the fracture abundance is, in my experience, often not well represented in standard cross plots of the data. While a trend is usually seen, probably because of the variance and complexity of the two data sets, correlation coefficients, or r^2 values, tend to be lower than considered conclusive. Examples of both good and bad production/fracture correlations on standard cross plots are shown in Figure 14.2a and b.

Alternatively, better representations of the correlation are more convincing displayed in map form or in individual well production profiles. Frequently, various forms of production characteristics are mapped in fractured reservoirs using "bubble map" formats, Figure 14.3. In that figure, cumulative liquid volumes are represented as bubbles or circles the radius of which is equivalent to the volume produced.

Alternatively, by defining areas or domains in map form that are dominated by high- and low-fracture intensity and correspondingly high- and low-PI values, for example, and overlaying the domain boundaries, we see very good spatial correlation. Figure 14.4 is one such example from the Middle East.

Additionally, similar map representations to show correlation can be done for parameters like drainage area per well, Figure 14.5.

This figure displays calculated circular drainage areas surrounding the wells. However, in most fracture-dominated fields, the drainage areas are highly elliptical, elongated parallel to the dominant natural-fracture direction or

Static Conceptual Fracture Modeling: Preparing for Simulation and Development,
First Edition. R.A. Nelson.
© 2020 John Wiley & Sons Ltd. Published 2020 by John Wiley & Sons Ltd.

Production Parameters to Try Correlating to
Fracture Intensity

1. Cumulative production
2. Percent of field-wide production per well
3. Initial Potential (IP)
4. Productivity Index (PI)
5. Maximum flow rate
6. Drainage area (area & shape)
7. Secondary recovery breakthrough (or short-circuit)
8. Wellbore skin
9. Wellbore interference data

Figure 14.1 Listing of production data or characteristics that can historically correlate well with natural Fracture Intensity.

(a)

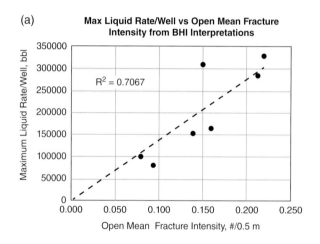

(b)

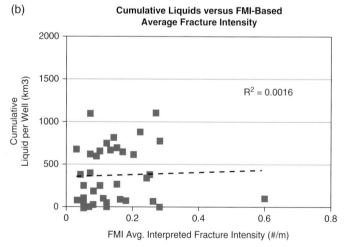

Figure 14.2(a and b) Cross plots of production data and observed Fracture Intensity are displayed. (a) shows an unusually good correlation in one field, and (b) a more typical weaker correlation in another.

Cumulative Production Bubble Map, Western U.S.

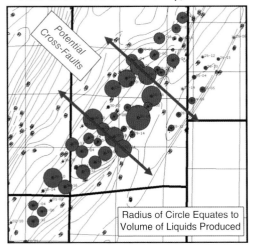

Radius of Circle Equates to
Volume of Liquids Produced

Figure 14.3 A bubble map of cumulative liquid production from a naturally fractured sandstone field in the western U.S. drilled with vertical wells. The variability seen is typical of fractured reservoirs with vertical wells as some intersect high fracture intensity corridors in hinges (right side of the field) and in cross faults (transverse alignments).

Example of Correlating Natural Fracture & Production Data in a Producing Middle East Oil Field by Mapping Parameter Domains

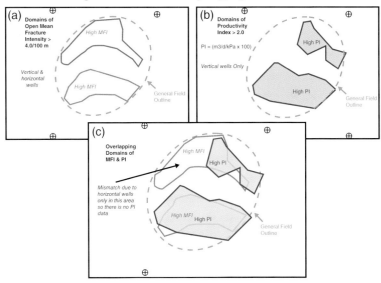

Figure 14.4 The parameters Mean Fracture Intensity (MFI) from BHI interpretations and Productivity Index (MFI and PI) were mapped and outlines of high values or high density of values drawn in (a) and (b). These domains were then overlain to show the spatial relationship between Average Fracture Intensity (AFI) and PI in (c). The non-overlay in the NW edge of the field is due to no PI data in those horizontal wells. These resulting domains show that the northern and southern portions of the field have the largest fracture effects, while the central portion has the least. These domains can be used to guide the distribution of discrete stochastic fracture corridors during future fracture modeling.

Calculated Drainage Radii Overlaying Fault Map

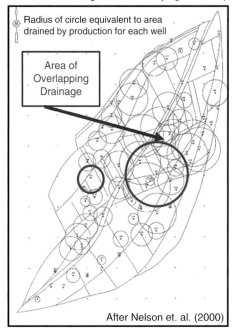

Radius of circle equivalent to area drained by production for each well

Area of Overlapping Drainage

After Nelson et. al. (2000)

Figure 14.5 Circular drainage area map for a large fractured carbonate field. Radius of circle is calculated from total liquids produced per well compared to the total porosity. Of importance is the area of highest production (east central part of the field) where all drainage areas overlap. One conclusion is that oil is recharging the structure from down dip along fracture corridors associated with faults in this area as production has taken place.

Mapped Elliptical Drainage Areas in Silo Field Western U.S. with Long Axis Parallel to Dominant Natural Fracture Direction

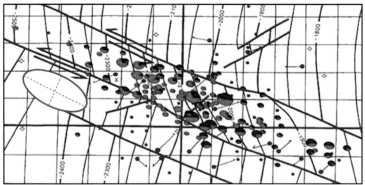

Figure 14.6 Example of measured or assumed elliptical drainage areas from the Silo Field in the western U.S. *Source:* after Sonnenberg and Weimer (1993). Courtesy The Mountain Geologist.

parallel to those fracture elements parallel to the horizontal maximum reservoir stress direction. Such a situation is shown in Figure 14.6 where calculated drainage areas are interpreted to be elliptical and parallel to the dominant natural-fracture direction.

15

Determining the Fractured Reservoir Classification and, Therefore, Which Simulation Style Is Most Appropriate

As a first pass, we need to understand the type of fractured reservoir we are trying to develop and model. By classifying the reservoir, we can determine several reservoir characteristics:

1) The parameters most important in quantifying the reservoir can be determined.
2) Potential production and evaluation problems can be anticipated.
3) Recovery factor and producible reserves can be estimated.
4) Inhomogeneity in production volumes can be predicted.
5) Rates can be estimated.
6) Reservoir permeability anisotropy can be predicted.
7) The style of reservoir simulation necessary can be constrained.

This classification is a simple one that uses the overall contribution the fracture system in total reservoir quality. The initial published concept behind this classification was by M. King Hubbert and D.G. Willis of Shell in the 1955 World Petroleum Congress Proceedings, Hubbert and Willis (1955). In that publication, they proposed two types of fractured reservoirs; one where the fracture system provides the effective reservoir storage and permeability, and a second where the fracture system provides the effective permeability and the matrix provides the storage.

This classification languished for some 30 years until it was modified and reintroduced in Nelson (1985), and later refined and detailed in Nelson (2001, 2002). In this expanded classification, there are four types of fractured reservoirs, the characteristics of which are detailed in Figure 15.1a. A graphic visualization of these types is shown plotting each type's position in percent of total porosity and percent of total permeability space in Figure 15.1b.

Once the classification of the fractured reservoir we are trying to model is determined, several reservoir aspects are envisioned. In one, production inhomogeneity can be predicted for the case of vertical well development. This will help constrain development drilling risk.

Static Conceptual Fracture Modeling: Preparing for Simulation and Development,
First Edition. R.A. Nelson.
© 2020 John Wiley & Sons Ltd. Published 2020 by John Wiley & Sons Ltd.

(a)   **Fractured Reservoir Classification**

> *Type I:* **Fractures provide the essential storage capacity and permeability** in a reservoir. The matrix has little porosity or permeability
>
> *Type II:* Rock matrix provides the essential storage capacity and **fractures provide the essential permeability** in a reservoir. The rock matrix has low permeability, but may have low, moderate, or even high porosity.
>
> *Type III:* **Fractures provide a permeability assist** in an already economically producible reservoir that has good matrix porosity and permeability.
>
> *Type IV:* **Fractures do not provide significant additional storage capacity or permeability** in an already producible reservoir, but instead create anisotropy. **(Barriers to Flow)**

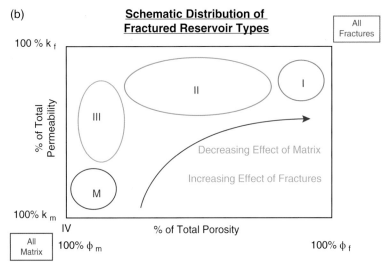

(b)   **Schematic Distribution of Fractured Reservoir Types**

Figure 15.1(a and b) (a) The characteristics of the four types of fractured reservoirs as defined in Nelson (1985, 2001), and (b) a schematic diagram showing the position of the four types in a percent permeability vs percent porosity plot. It shows the trend of increasing effect of fractures in a manner quite familiar to reservoir engineers. *Source:* courtesy of Butterworth Heinemann.

In Nelson (1985, p. 115) a technique was first described to investigate reservoir inhomogeneity in terms of both cumulative production and maximum production rate. To create a graphical representation of the cumulative production inhomogeneity, we order the producing wells from the lowest cumulative production to the highest. We then plot the ordered wells on the x-axis and on the y-axis plot the percent of the total field-cumulative production that the well

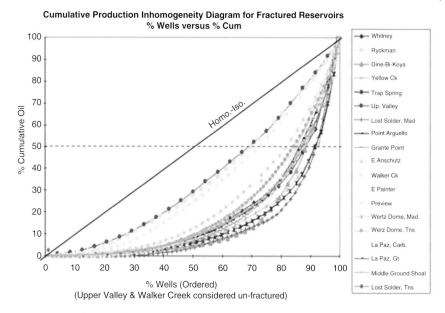

Figure 15.2 A fractured reservoir inhomogeneity diagram as described in Nelson (2001). It plots data from vertical well production in a field against the ordered production contribution of each well, smallest to largest. The history of 17 fractured reservoirs and two matrix-only reservoirs are plotted on the diagram. If every well in the field produced the same cumulative production, the curves would be a straight 45° line. Due to stratigraphic inhomogeneity and inconsistences in flow control and completions, no field would meet this line. However, the fractured reservoirs plotted show much more inhomogeneity in cumulative production, a characteristic of fractured reservoirs. The red dashed line indicates a 50% cumulative production line. The number of wells above the intersection of the field curve and the 50% line means that 50% of the fieldwide cumulative productions came from those few remaining wells. *Source:* figure from Nelson (2001) courtesy of Butterworth-Heinemann.

contributed to the total field-wide volume, Figure 15.2. If every well produced the same volume, the generated curve would be a straight 45° line, showing homogeneous production. As seen in this figure, the resultant curves for several fractured reservoirs differ significantly from the homogeneity line, being displaced to the lower right quadrant of the plot. The amount of displacement to the lower-right quadrant is termed the "Fracture Impact Coefficient (FIC)," or more aptly an "Inhomogeneity Coefficient" (IC) and is calculated as the percentage of area of deviation of the resultant curve to the lower-right quadrant below the 45° line. The range in IC observed in that study of various host rock types is 0.28–0.73 with a possible range from 0 to 1. This IC is similar in concept to a "Lorenz Coefficient" in reservoir engineering which also looks a reservoir heterogeneity, mostly from a stratigraphic point of view. That approach was originally developed by others as a tool in economic analysis and later applied to reservoir engineering.

The Fractured Reservoir Classification appears to discriminate between fracture reservoirs by the degree of displacement of the curves to the lower-right quadrant, Figure 15.3.

Another way the Fractured Reservoir Classification impacts Static Conceptual Fracture Model (SCFM) modeling is in the most appropriate type of reservoir simulation to be applied. There are three basic styles of simulation that can and are used in simulation of fractured reservoirs. These are shown in Figure 15.4, and include Single Porosity, Dual Porosity, and Dual Porosity-Dual Permeability simulations; the physical definition of each is included.

Applying these definitions to the percent porosity–percent permeability plot format, we see that the Fractured Reservoir Classification tells us which simulation style is **most appropriate** to use for any specific fractured reservoir type, Figure 15.5. However, what usually happens is that our initial models are often poorly constrained by data and, by default, the simpler single-porosity models are used initially. With time and additional reservoir and flow data, the more appropriate and complex modeling styles will be attempted to better constrain the reservoir.

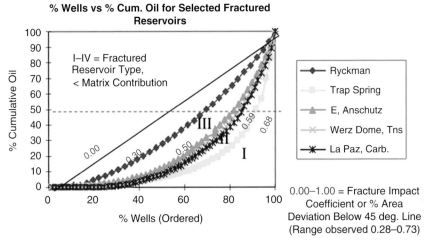

Curve position a function of fractured reservoir type

Figure 15.3 An inhomogeneity plot, as described in the previous figure, with curves from five fractured reservoirs that represent different classes of fractured reservoirs as described in Nelson (2001). It is evident that Type III fractured reservoirs show a small degree of production inhomogeneity. Types II and I fractured reservoirs display progressively more inhomogeneity. More inhomogeneity means that there will be greater differences in cumulative production from the wells. The amount of displacement of the curves to the lower right (percentage of quadrant area displaced) is called a Fracture Impact Coefficient (FIC), or better called an Inhomogeneity Coefficient (IC) with the higher percentage indicating more inhomogeneity in production. *Source:* courtesy of Butterworth-Heinemann.

Styles of Reservoir Simulation in Fractured Reservoirs

- **Single Porosity**
  - 1 medium for fluid flow.
  - Can be matrix-only, fractures-only, or matrix plus fractures added together as a single medium with hybrid flow properties.
- **Dual Porosity**
  - 2 Media for fluid flow (matrix and fracture flow properties).
  - Fractures flow to other fractures.
  - Matrix blocks can flow only to fractures (matrix blocks cannot flow to other matrix blocks directly)
- **Dual Porosity- Dual Permeability**
  - 2 media for fluid flow (matrix and fracture flow properties).
  - Fractures flow to other fractures.
  - Matrix blocks flow to fractures.
  - Matrix blocks flow directly to other matrix blocks

Figure 15.4 Listed are the three basic types of simulation style used in simulating fractured reservoirs. They range from the simple to the complex and from small data requirements to large. Included are definitions of the physics of the individual styles.

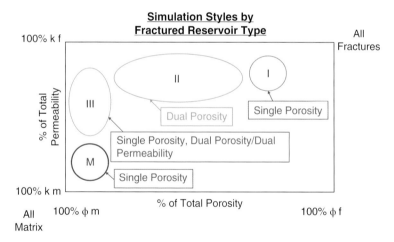

Figure 15.5 Given the definitions of simulation styles in fractured reservoirs shown in Figure 15.4, this figure highlights which style is most appropriate for each reservoir classification in a percentage permeability/percentage porosity format. *Source:* figure is from Nelson (2001).

Therefore, if we can classify the fractured reservoir type that we are going to model, it will tell us how to predict production problems, the most important modeling parameters, production inhomogeneity, proper simulation style, possible rates, reserves and recovery factor, (Nelson 2001).

16

Use of Reservoir Analogs

To create a Static Conceptual Fracture Model (SCFM) for a reservoir, we measure many parameters individually and use them as constraint in our modeling. However, as was pointed out previously, we often lack first-hand data on many of those parameters, particularly at the beginning of a project. To fill this need we frequently utilize fractured reservoir analogs of producing or produced fields that share characteristics with the field we are modeling. This data can come from published sources or from our collective personal experience as analysts. These analogs can fill in the gaps in our databases, help predict evaluation or production problems, and anticipate sensitivity of various parameters in the modeling process. The analogs we choose can be individual parameters that are well-constrained in another, but similar field, or groups of parameters from a field. Also, individual parameters, like fracture spacing or fracture aperture, can be utilized directly as values or used as a form of distribution function applied to whatever field specific data exists in the field being modeled.

There are a variety of reservoir analog parameters that we can call on, each applying to a specific part of the process. These individual analogous parameters include those listed in Figure 16.1.

As we start to define our reservoir for modeling, we look for other old or producing fields that can provide a context for our reservoir of interest. For example, if we were modeling a limestone reservoir with 2% average matrix porosity in a strike-slip related fold with only vertical wells and no hydraulic fracture stimulation we might use parts of La Paz Field in the Maracaibo Basin of Venezuela. We can then look at other aspects of the field's history and production that might be relevant to our modeled field, for example, other analog types from that chosen field.

These analog types shown in Figure 16.1 can be gathered together in a compilation of fractured reservoirs in spreadsheet format that can be sorted in different ways to find these important analogs. I have created one such database over the years and use it to jump-start many a model. Mine has quantitative reservoir and production data for 130 fractured reservoirs but can, and will

Static Conceptual Fracture Modeling: Preparing for Simulation and Development,
First Edition. R.A. Nelson.
© 2020 John Wiley & Sons Ltd. Published 2020 by John Wiley & Sons Ltd.

Possible Types of Reservoir Analogs
Production
Rate
Behavior
Well Type
Completion Type
Stimulation Type
Volume
Development Style
Rock Type
Fluid Type
Depth
Reservoir Properties
Subsurface Structure
Structural Style
Structural Sequence
Reservoir Mechanics
Historical
Regional and/or Basinal
Stress State
Diagenetic History

Figure 16.1 Listing of possible analog types that can be useful in modeling a fractured reservoir, particularly in the early stages of modeling.

undoubtedly, be expanded in the future. This analog database is constructed in spreadsheet format and has bins for the following data on each field, Figure 16.2.

The beauty of the spreadsheet format is that you can sort the fields by various parameters like lithology, formation name, or matrix properties. In fact, most spreadsheets allow you to sort on multiple parameters at the same time, making the process of analog selection easier.

Several geological service companies have created their own inhouse Fractured Reservoir Analog databases that can be used to find characteristics of a reservoir like the one you are modeling. All record slightly different field parameters and contain different levels of source referencing.

Application of fractured reservoir analogs can be extremely important in filling in the gaps in our quantitative database and in predicting reservoir behavior with time, which after all is the purpose of the eventual reservoir simulation. However, an important aspect of these analogs early in the process is in getting company management on board with exploration and development plans for the reservoir we are modeling. We need to show a real-life example of a field

Analog Parameters
Field Name
Location by Country
Reservoir Age
Reservoir Formation Name
Lithology
Environment of Deposition
Fracture Origin Type(s)
Fractured Reservoir Types (1-4)
Fracture Spacing
Matrix Porosity
Fracture Porosity
Matrix Permeability
Fracture Permeability
Fluid Type
STOIP
Recovery Percent
API Gravity
Number of Wells in the Field

Figure 16.2 Listing of reservoir parameters useful to include in a Fractured Reservoir Analog spreadsheet to aid in the SCFM process.

that looks like ours and points a way toward economic development. This includes things that were tried that worked or didn't work and things that could have been done better if more were known at the time. This anecdotal history can prepare us for unpredicted outcomes along the way.

17

The Importance of 3D Visualization in Data Integration and Static Fracture Model Creation

In modeling fractured reservoirs and the creation of a robust Static Conceptual Fracture Model (SCFM), we must integrate many individual data sets from multiple disciplines. Creation of these data sets is best initiated and facilitated by a Fracture Study Champion. The champion can be a manager, team member, or external consultant whose role is to foster creation of the needed data sets, facilitate the workflow, make sure the format of the data and interpretations is conducive to integration, and guide the multidisciplinary team members through the data integration process. The critical integration process should take place toward the end of the modeling process after all data is loaded into the available computer-modeling software. What is imperative in this integration is the ability for all team members to see their data shown in context with that of all other team members. In my experience, this integration is best accomplished using 3D visualization.

Most fractured reservoir modeling programs (in my experience Petrel, FracMan, and other in-company proprietary programs) allow for multiple data sets to be co-rendered in 3D with the ability to rapidly switch between alternative views of the data in real time. I believe that this 3D visualization performed in a team integration context is the best and most efficient way to integrate the data and make high-level correlations and interpretations. To do this, all technical team members, and only technical team members, including the Fracture Champion should be in attendance in the integration sessions. Integration could take place over multiple sessions with adjustments to the database and alternative visuals generated between sessions at the request of the team. The Fracture Champion, and the person in charge of the computer modeling software should manage the session and the team members should drive the various displays and the direction of the integration. Indeed, team members should request alternative displays of data in real time to effectively accomplish innovative integration. An example of the kind of displays helpful in this process is given in Figure 17.1.

Static Conceptual Fracture Modeling: Preparing for Simulation and Development,
First Edition. R.A. Nelson.
© 2020 John Wiley & Sons Ltd. Published 2020 by John Wiley & Sons Ltd.

**Fracture Intensity, Porosity, Mapped Faults, &
Interpreted Coherency Features**

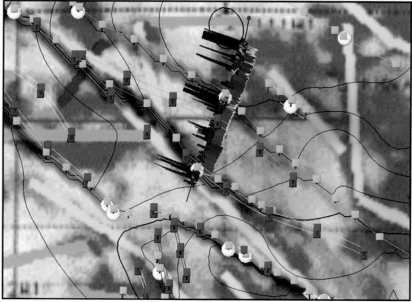

Figure 17.1 An example of 3D data-integration display used in an SCFM integration meeting. This example is from a producing field in the Middle East. Included is: a seismic coherency map as a base with interpreted coherency-based faults in yellow, red and green differentiated by azimuth class, 3D seismic mapped faults in pin lines, structure contours in black, and a horizontal wellbore with porosity curve and FI curve. Notice the exceptional consistency between the data types. This type of display fosters integration and buy-in from all technical team members.

In my experience during integration sections for SCFM creation, some of the requested views that naturally surface include the following:

1) Fracture Intensity (FI) from wellbores plotted along wellbore length along with measured mechanical properties to derive a mechanical explanation for FI distribution and a resultant predictor of distribution in less constrained areas.

2) Percentage of Cemented Fractures plotted on structural horizon maps to envision where diagenetic fluids are entering the reservoir, such as along mapped faults or fold hinges. If seen, this can lead to predicting areas of higher open-fracture permeability and storage.

3) FI measured in wellbores plotted with standard wireline log response and interpreted stratigraphy that the team is using. In displaying the log curves (FI and wireline) care should be taken to ensure that proper

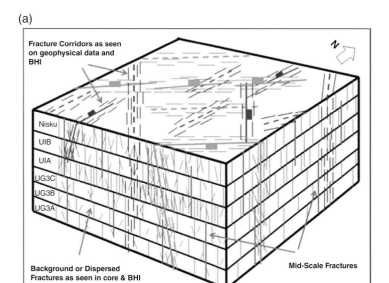

Tectonic fractures, normal fault state of stress, 3 periods of deformation, each period of deformation has unique fracture properties (azi., spacing, length, aperture), all span multiple scales of development

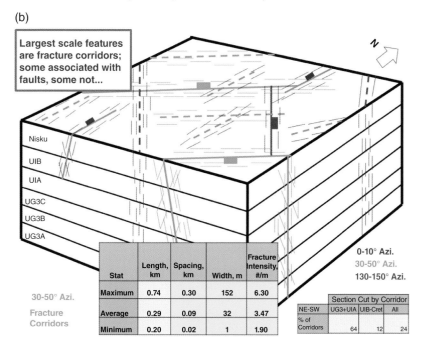

Figure 17.2(a and b) Schematic block diagrams representing the Grosmont Reservoir in Alberta, Canada. Diagrams like these are critical for sharing the qualitative and quantitative data and interpretations in an SCFM with all members of the technical team, as well as management. In (a) we see all elements of the multi-scale fracture system including background fractures, mid-scale fractures and fracture corridors, and (b) a statistical representation of just one element of the system, the fracture corridors. *Source:* figure from Wagner et al. (2010).

sensitivity and smoothing is applied to the FIC to more easily compare the various curve data.

4) Also, I have seen instances in the integration where differences arise between interpreted unit tops, which are usually placed at Gamma Ray Log changes, and FI and/or mechanical property interpreted tops. In several cases, the interpreting team has changed the interpreted stratigraphic tops for mapping and reservoir architecture to the mechanical one.

An output from the team integrations session is usually one, or a set of, block diagram schematics that depict the data distribution and ranges of parameters visually so that all members have a common visual understanding of the current view of the model, Figure 17.2a and b.

Block diagram schematics for the SCFM can be constructed (Figure 17.2a) and used as a base to post the spacing and width statistics for each of the different scales of interpreted and measured faults in an area (Figure 17.2b).

18

Thoughts on History Matching of Simulation Results

After one or more data integration sessions, the Fracture Team creates a gridded 3D Static Conceptual Fracture Model (SCFM) to feed the reservoir simulation.

History matching occurs when the initial flow simulation is done, and the results are matched to actual production, well test results, or reservoir pressure profiles. Alterations are then made to the input data (SCFM or Dynamic Conceptual Fracture Model (DCFM)) and re-run to achieve a better match of the resultant simulation to control points in the reservoir. However, even in the larger E&P companies, these alterations are generally made by the reservoir engineer doing the simulation, and not the team that created the SCFM. This is inappropriate. The team that constructed the SCFM knows the relative strengths and weaknesses of the input data and are the only ones with the knowledge of which parameters should be changed and how to change them to assist the simulation modeler in response to history matching. I have seen this process go very wrong at this stage in companies both large and small.

Another problem that I have seen during history matching is inappropriate no-flow boundaries in the initial simulation model. In one modeling effort in the Middle East, the top of simulation was what was called a "shale" in the stratigraphic section due to a higher gamma-ray response on the logs. As such, it was modeled as a no-flow boundary to the simulation. In reality, it was a fine-grained, low-porosity carbonate unit that was mechanically the stiffest unit in the reservoir section. The result was that it contained the highest natural-fracture intensity in the reservoir section as eventually confirmed by further analysis of the Formation Micro Imager (FMI) fracture interpretations. Initial simulations were history matched and all changes to the SCFM were made in the reservoir below this unit. Later, Repeat Formation Tester response proved that the upper zones were pressure depleted due to production and the simulation was really an open-flow system and not a closed one, invalidating the results of the simulation.

Static Conceptual Fracture Modeling: Preparing for Simulation and Development,
First Edition. R.A. Nelson.
© 2020 John Wiley & Sons Ltd. Published 2020 by John Wiley & Sons Ltd.

19

Preparing the Fracture Data for Input to the Gridded Model

Once we have studied the natural-fracture parameters needed to constitute the Static Conceptual Fracture Model (SCFM) in as complete manner as the available data allows, it is time to put spatial and quantitative data into the formats needed to create the final gridded fracture model that will go into the reservoir simulator. This includes placement of features (like fracture corridors and faults) into the model and either the distribution functions or statistical representations of the parameters needed to in the final model.

As stated previously, our modeling choices include non-discrete and discrete and within the discrete, deterministic and stochastic. Recall that in non-discrete modeling we approximate the natural fracture system by adjusting the porosity and directional permeability of the matrix to respond to production, like fractures were there. However, the purpose of creating an SCFM is to model the reservoir for simulation in a discrete modeling fashion (a Discrete Fracture Network or DFN). Therefore, the remainder of this section will focus on generating discrete deterministic and discrete stochastic modeling aspects of the use of the SCFM.

In discrete stochastic modeling of a fractured reservoir, we have the computer place localized and distributed fracture features statistically within the reservoir volume. As we generally model more than one scale of fractures from background to corridors and faults, the location, orientation, spacing and intensity of the fractures and features are placed in the model numerically, guided completely by the statistics of the parameter populations. Therefore, it is appropriate to organize the parameter data into as full a distribution as possible. This is done by either measuring a nearly complete distribution and generating a "best-fit" to the data points in a program like Excel or a similar curve fitting program, or by taking a more limited set of measurements and applying them to "standard distribution functions" known to fit natural-fracture parameter distributions. While there are numerous statistical curve types, in my experience as a fracture consultant and not a statistician, those that are most often used in the fracture modeling literature to fit sparse fracture data sets are either normal, power-law, and log-normal distributions, Figure 19.1.

Static Conceptual Fracture Modeling: Preparing for Simulation and Development,
First Edition. R.A. Nelson.
© 2020 John Wiley & Sons Ltd. Published 2020 by John Wiley & Sons Ltd.

**Standard Distribution Function Curve Shapes Often
Used in Natural Fracture Modeling**

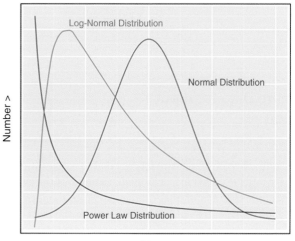

Figure 19.1 The three basic forms of parameter distribution curves most often used in natural fracture modeling are shown. They are a Normal, or symmetric, Distribution, a Power-Law, or hyperbolic, Distribution, and a Log-Normal, or highly-skewed, Distribution.

In addition, the power-law distributions can be displayed as either nonlinear (standard) or linear (fractal) relations, Figure 19.2. If dictated by the measurements, the distributions can be uni-modal or poly-modal in shape and can display some degree of skewness. Modeling parameters may each conform to different distribution functions. The amount of data and the physics of the generative process control which of these distributions are most appropriate for an individual parameter. For example, scalable parameters such as fracture length and spacing have been shown to follow Log-Normal and Power-Law distributions, including fractal relations; Qinghaul, et al. (2017), Gutierrez and Youn (2015), Massiot et al. (2015), Wagner et al. (2011) Davy et al. (2018), Bisdom et al. (2014). Fracture aperture has been shown to follow either Log-Normal or Power-Law Fractal relations; Marrett et al. (1999), Gale (2014), Qinghaul et al. (2017), Massiot et al. (2015). Fracture-set orientation generally follows a Normal Distribution, although if multiple fracture sets are present, the symmetry may be altered to a poly-modal distribution.

In some fracture modeling software, we need to input the parameter data not as curves but in statistical format (for example maximum, minimum, mean, most likely, and standard deviation) and the software will apply the appropriate shape of the distribution curve. The same software we would use to best-fit our data can also generate the required population statistics.

Two Types of Power Law Distribution Plots

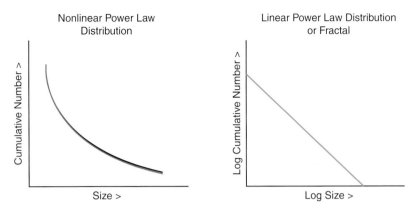

Figure 19.2 There are two types of Power-Law Distribution display formats. One is nonlinear and is a hyperbolic form and the other is a linear form as plotted on a log–log plot showing the self-similar nature of the fractal. *Source:* after Cioffi-Revilla and Midlarsky (2013).

For discrete deterministic modeling, for all but the background fractures, we place the features into the model where we know they are in the reservoir. We can know these positions for several reasons; (i) we documented the features within our wellbores, (ii) we mapped them in our surface data or in 3D geophysical data or, (iii) we have seen indications of their presence from production and drilling response. For example, in modeling software like Petrel we can define features like fracture corridors and faults directly on well logs in well sections and condition the modeling to stochastically place features of known properties at that well location or, conversely, exclude one from being placed there because no evidence was seen.

20

Discussion of Error and Uncertainty in the Modeling Process

Of course, there is error and uncertainty involved in the process of creating a Static Conceptual Fracture Model (SCFM). After searching the literature, various dictionaries, and Wikipedia for these two terms, I chose the following definitions as they apply to this modeling process:

Error indicates the confidence level in a physical measurement or inaccuracy or lack of sureness.

Uncertainty is a situation which involves imperfect or unknown information or doubt. It applies to predictions of future events, to physical measurements that are already made, or to the unknown.

Both apply to the process of creating an SCFM and the reservoir simulation that it feeds. One presentation that details uncertainty in the process is Fox et al. (2009). In that presentation, is a breakdown of individual modeling parameters that are associated with different classifications of uncertainty.

As part of the fracture modeling process, we derive measured parameters and create the distributions of those parameters (fracture orientation, length, spacing, aperture, etc.). The associated measurement error involves several factors, Table 20.1.

There is a precision or +/− error associated with the instruments or tools we use to constrain a variable. In our modeling work, our tools can vary substantially when measuring a single variable. For example, we can measure the width of a fracture corridor by its width on a Borehole Image (BHI) interpretation in a horizontal well. However, as we have stated previously, not having a horizontal wellbore with a BHI log may lead us to constrain fracture corridor width by interpreting seismic attribute maps, like positive and negative curvature maps. The precision of the two data sets are vastly different. Similarly, fracture aperture measurement made by using comparators (lines of fixed known width) at the core will be quite different than those obtained from BHI resistivity logs. Therefore, the type of tool we use in measurement dictates the precision error we will have.

Static Conceptual Fracture Modeling: Preparing for Simulation and Development, First Edition. R.A. Nelson.

Table 20.1 These are the types of error involved in measurement error of modeling parameters.

Sources of Error of Parameters
1) Precision of the measurement instrument or tool
2) Statistics of the fit of the distribution function
3) Operator variability
4) Propagation of measurement errors through the calculation function using several parameters

---

Once we make physical measurements of a parameter the modeling process requires that we generate an appropriate parameter distribution function. There is additional error associated with "fitting" the measurements to a distribution function as measured by something like a standard deviation or an r^2 value.

An additional source of error can be operator variability. As shown in Nelson et al. (1987) having multiple interpreters interjects additional variability in measurement. A prime example is in fracture interpretations from core and from BHI images. There, operator variability can be huge. This is one reason it is recommended that either one or a limited number of interpreters interpret all cores and BIL within a field study.

Lastly, these parameters with their individual errors are plugged into equations using multiple variables and the errors of each are propagated through the calculations. Paraphrasing a quote by M. Palmer from MIT online "Since a property of interest in an experiment or model is rarely obtained by measuring it directly but by calculation from other measured values using an equation, we must understand how error in direct measurements propagates when mathematical operations are performed on these measured quantities."

All in all, measurement error can be very large in creation of an SCFM and can vary significantly from project to project depending on the data we have available and the tools we choose to quantify them.

Uncertainty is also quite large in fracture models leading to Discrete Fracture Network (DFN) creation. Doubt and lack of sureness is high when we do not even have data on certain parameters or the data that we do have is sparse, i.e. few wells with core or appropriate logs, lack of wells in large portions of the field, or lack of 3D seismic surveys.

Because we are usually working on fractured reservoirs with relatively large errors and uncertainty due to the factors mentioned above, modelers often make many realizations of the model in a Monte Carlo style to gain knowledge of sensitivity to the various parameters and constrain uncertainty.

21

Published Examples of the Process

There are several papers that, in principle, discuss the process that is outlined in this book. Indeed, several excellent examples include steps not included in this book but are very effective additions. In each, the steps are not identical to what is included here, but the overall flow to the work is. The first good example is found in Richard et al. (2017). That paper is authored by a collaboration of workers from Shell and the Kuwait Oil Company (KOC). It is an excellent example of the breadth of a Static Conceptual Fracture Model (SCFM) project taken to simulation.

As stated in that manuscript, "This paper illustrates some examples of best practices of the various study components with a focus on core to BHI calibration, fracture-porosity calibration using core data and calibration of DFN models using pressure transient analysis data." Their successful approach relied particularly on core, image log interpretations, and analysis of structural development and sequencing. This was aided by use of a proprietary Shell software package called FFS – Fault and Fracture Solutions. The reliance on the structural sequencing approach and the in-house software is what separates the Shell/KOC approach from that detailed in this book. An example of the structural sequencing used in that paper, but not covered in this book, is shown in Figure 21.1.

In this approach, each time compartment is ascribed a different level of deformational strain that is then used to predict fracture intensity and other fracture properties during that time period. This fracture generation modeling is an expansion to what is covered in this book.

Another excellent published example of a similar type of modeling process is Paul et al. (2007). That work also uses mechanical modeling but this time to generate fault-zone properties and fractures in the model. They borrow a process from earthquake seismology called Dynamic Rupture Propagation and apply it to the SCFM process. The same elements we discuss in this book are then used to calibrate the results from the mechanical prediction. The fault zone that they model is in the CS Field in the Timor Gap between Australia

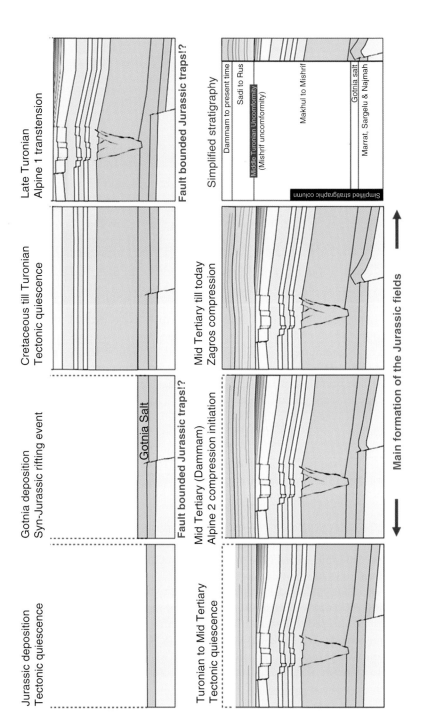

Figure 21.1 An example of structural sequencing used in creation of an SCFM in Kuwait. *Source:* from Richard et al. (2017), courtesy of Society of Petroleum Engineers.

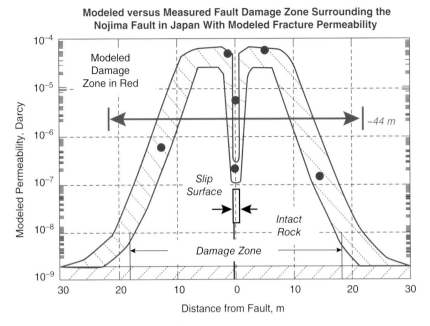

Modeled versus Measured Fault Damage Zone Surrounding the Nojima Fault in Japan With Modeled Fracture Permeability

Figure 21.2 One example of a modeled damage zone surrounding a fault in crystalline rocks after Paul et al. (2007). The mechanical model is generated using Dynamic Rupture Propagation that is then calibrated using many of the same static parameters presented in this text. *Source:* courtesy of Society of Petroleum Engineers.

and Indonesia. Their purpose was to define the width and permeability of the "damage zone" (width of deformation surrounding the fault-slip surface). One example from their modeling is shown in Figure 21.2.

The two examples included in this section both include mechanical modeling steps but of different types. I believe that this is the future of SCFM processes.

22

Final Comments

What we choose to use to constrain the variable parameter distributions for Static Conceptual Fracture Model (SCFM) creation, is totally dependent on the reservoir and subsurface data we have at the time. It is important to understand previous work in the area, know the state of the art in obtaining data and then to approach the modeling in a manner that is **innovative rather that rote**. In SCFM generation, we must attempt to constrain the input variables as best we can whether it is by measurements, by correlation to other variables, or by analog. This book can be a guide to the important parameters needed to construct an SCFM and the various techniques we can use to gather data on constraining those parameters. In addition, insights are given as to the challenges involved with applying those techniques and the potential error involved in their measurement. The bottom line is a path to the data gathering, integration, and data preparation needed to construct a good Static Conceptual Fracture Model.

Static Conceptual Fracture Modeling: Preparing for Simulation and Development,
First Edition. R.A. Nelson.
© 2020 John Wiley & Sons Ltd. Published 2020 by John Wiley & Sons Ltd.

Appendix A

Detailed Static Fracture Modeling Workflow

The following text refers to Figure A.1, a schematic flow diagram organized by timing of the work, and Figure A.2, a schematic flow diagram organized by discipline focus or ownership.

1) **Gather Fracture/Reservoir Screening Data and Determine if it Should be Modeled as a Fractured and/or Unconventional Reservoir**
   a) **Determine if it is a fractured reservoir**
      i) Obtain matrix P&P data.
      ii) Obtain core and relevant outcrop fracture observations and determine the degree of fracturing.
      iii) Gather well-analyzed flow test or drill stem test type (DST) data and horizontal permeability (Kh) and compare to core analyses (if much greater than core, yes, it is a fractured reservoir.
      iv) Determine reservoir flow anisotropy (if > 10, yes, it is a fractured reservoir)
      v) Match reservoir Kh to core based Kh (if Kh test > Kh core, if yes it is a fractured reservoir)
      vi) See also Nelson (2001).
2) **Gather Relevant Structural Geology Input**
   a) Well-constrained structure maps at all relevant reservoir horizons
   b) Significant seismic attributes that constrain structural features (i.e. coherency, ant tracking on discontinuity data, structural curvature).
3) **Determine Fractured Reservoir Type**
   a) *Type I*: Fractures provide the essential storage capacity and permeability in a reservoir. The matrix has little porosity or permeability.
   b) *Type II*: Rock matrix provides the essential storage capacity and fractures provide the essential permeability in a reservoir. The rock matrix has low permeability, but may have low, moderate, or even high porosity.
   c) *Type III*: Fractures provide a permeability assist in an already economically producible reservoir that has good matrix porosity and permeability.

Static Conceptual Fracture Modeling: Preparing for Simulation and Development,
First Edition. R.A. Nelson.
© 2020 John Wiley & Sons Ltd. Published 2020 by John Wiley & Sons Ltd.

d) *Type IV*: Fractures do not provide significant additional storage capacity or permeability in an already producible reservoir, but instead create anisotropy. (Barriers to Flow).

4) **Fracture Origin**
   a) Review structural mapping for folding and faulting (trends, axes, traces)
   b) Review outcrop patterns for structure and fractures
   c) Analyze strike and dip patterns considering standard templates of fracture origin (Nelson 2001).

5) **Fracture Orientation(s) and its' distribution**
   a) Perform lineation analysis
      i) At surface (remote sensing)
      ii) At reservoir (structural mapping)
      iii) At basement (seismic and/or potential fields).
   b) Interpret fracture orientation from relevant outcrops
      i) Determine real from induced fractures.
   c) Interpret fracture orientation from Borehole Image Logs (BHI) and/or core
      1) Determine scale and resolution of data source
         a) Core: core slab photos, core slabs, whole core, slabs + butts
         b) Image Log: Which log type, which service Company, which operator?
         c) Core and Image Log together: How many interpreters?
         d) Determine real from induced fractures.
      2) QC Borehole Image Quality and core condition and recovery
      3) QC fracture interpretations from vendors and other operators, if appropriate.
   d) Interpret single-dominant open fracture orientation from borehole seismic anisotropy in shear sonic log data
   e) Map fracture orientations in map distribution across the area of study from all data sources annotating data source at each well
   f) Map fracture orientations in log form in individual wellbores from BHI and/or core in strip log format.

6) **Fracture Intensity or Spacing and its' distribution**
   a) Direct Observation
      i) Map fracture intensity from relevant outcrops in 3D. Measures should be taken in single units across the structure(s) sampling various strain zones across the structure and in single locations in varying stratigraphic/mechanical units. Vertical fracture intensity should be measured and concurrent horizontal fracture spacing in the same location(s) by azimuth to allow for turning vertical fracture intercept rates in vertical wellbores into horizontal fracture spacing.
      ii) Calculate fracture intensity in wellbores form BHI logs and core
         1) In at least one instance, analyze a core from the same interval in a well as on an image log. If more than one image log type is used for fracture interpretation (resistivity, ultrasonic, visual, etc.) interpret at least one instance of analyzing core and all image logs

used in a common well. This allows for calibration of multiple data type for future instances from a single data type.

iii) Create Fracture Intensity Curves (FIC); vertical, horizontal, or deviated fracture intercept rates. Create for "All Fractures," "Filled Fractures," and "Open & Partially Open Fractures."

iv) Geometrically correct natural fracture intercept rates for sampling bias due to well bore azimuth and fracture set azimuth if needed for horizontal or highly deviated wellbores.

v) Calculate Fracture Intensity values for each well for the entire measured well and for each important reservoir unit within the well separately. Do this for "All Fractures," "Filled Fractures," and "Open & Partially Open Fractures."

vi) Load FIC's in well log format into a 3D visualization package, such as Petrel™, Earth Vision, GeoProb, Rock Works, or other package used for fracture modeling.

vii) Create Percentage of Open and Partially Open Fractures values for each well and for each stratigraphic/mechanical unit within the wells.

viii) Create a table of the above of all fracture intensity data for all outcrops, all wells and all units. Create histograms of fracture intensities between wells and averaged by unit over the area of study.

ix) Create Fracture Intensity (or Fracture Spacing) maps and percentage Open and Partially Open Fracture maps for each of the wells and for each stratigraphic/mechanical unit within the wells

b) Construct a Mechanical Stratigraphy or Deformation Stratigraphy for the rock section of interest.

i) Map differences in deformation (fracture amount, fracture morphology, pressure solution, ductile flow) in cores and/or outcrop. Define mechanism of deformation behavior vertically in the rock section.

ii) Define the mechanical properties in the rock section by plotting rock mechanics moduli through the section. This is best done via dynamic moduli from sonic log data.

1) Plot logs of Rigidity or Shear Modulus (G) ($G = E/2(1 + \gamma)$) from shear wave sonic log data (Dipole Shear Sonic Imager (DSI), Sonic Scanner (SSCAN), etc.). In practice ($G = \rho(Vs^2)$ and load as logs at wells into the 3D visualization package, or

2) Plot logs of P-Wave Modulus (M) from compression wave sonic log data ($M = \rho[Vp^2]$). Correlate to G in one data set and correct to a Pseudo-G curve for wells that do not have shear sonic data. Load at wells into the 3D visualization package.

c) Indirect Observation

i) Select seismic attributes to quantify fracture intensity distributions from folding, faulting and large-scale fracture corridors

1) There are 17 seismic attribute techniques that have been used to document fracture distribution, see Attributes Chart

2) Use only attributes for which the physics of why they should detect volumes of elevated fracture intensity is easily understood

3) Highest confidence of results should be where zones of elevated fracture intensity are seen on multiple attribute displays

ii) Selected attributes must be calibrated against direct observation data on fracture distribution at surface or in wellbores. Attribute maps can then be turned in quantitative fracture prediction maps with the intensity within and the potential size or limits of anomalies quantified.

iii) Create anomaly domains and lineation and place on structure map and annotate the number of attributes that support the domains or lineation.

iv) Perform a statistical analysis of attribute anomalies to determine over what percentage of the mapped structure they occur (% of mapped closure, % of fault length or trace over which the anomaly occurs, distance from faults or fold hinge).

v) Record the variation in size and spacing of all selected domains.

7) **Fracture Morphology/Fill**

a) Record fracture morphology or fracture fill in outcrop and core using the description detailed in Nelson (2001). This includes:

i) Open Fractures

ii) Filled or Mineralized Fractures

iii) Incompletely Filled or Mineralized Fractures (Partially Open)

iv) Deformed Fractures

1) Slickensided

2) Deformation Bands.

b) Record fracture morphology or fracture fills from BHI logs. These will be called different things by service companies and is also dependent on the type of BHI interpreted. Fracture Morphology often limited on the BHI's.

i) Resistivity Image Log (e.g. Formation Micro Imager Log (FMI), Compact Micro Imager (CMI), Extended Range Micro Imager (XRMI))

1) Conductive Fracture (Open Fracture)

2) Resistive Fracture (Filled Fracture)

3) Partially Conductive Fracture (Partially Open Fracture)

4) The modifier of Continuous and Discontinuous is used to distinguish complete versus partial fracture traces on the image.

ii) Ultrasonic Image Log (e.g. Ultrasound Borhole Image Log (UBI))

1) Open Fractures

2) Partially Open Fractures

3) Possible Filled Fractures: If filled with minerals are acoustically different than the country rock (seen on the amplitude not time image).

c) Define fracture/matrix interaction (ability for cross flow) from direct observation of fracture surfaces from core and outcrop samples
   i) Hand samples
   ii) Thin sections
   iii) Oriented P&P plugs parallel and perpendicular to natural fractures.

8) **Selection of static conceptual fracture modeling style**
   a) There are several styles or approaches of Static Conceptual Fracture Model (SCFM) construction. Most include multiple scales of natural fracture development. Choice of modeling style should be made based on interpreted origin(s) of the fracture system present. Origin-related templates for fracture distribution and modeling include:
      i) Regional Fractures – two scales of development, relatively constant fracture strike, strong bed-contained and mechanical-unit containment, strong consistency in orientation across the area.
      ii) Fold-Related Fractures – three scales of development, complex orientation pattern with six dominant fracture set azimuths while being perpendicular to bedding, occurs with variable intensity everywhere on the fold.
      iii) Fault-Related Fractures – two or three scales of development, restricted to zone adjacent to the fault surface, parallel fracture traces with more complex dip patterns, zones can cross cut and intersect multiple units
      iv) Exposure-Related Fractures – Various fracture sizes restricted to exposure surfaces (erosional, diagenetic), discontinuous in vertical and lateral extent, wide range of apertures.
      v) Contractional Fractures – one scale of development, high degree of connectivity, low-permeability anisotropy, restricted to individual rock units.

9) **Fracture Aperture**
   a) Measure the mechanical fracture aperture of fractures in outcrop and/or core. Ideally, this is done on open and filled fractures separately with maximum, minimum and average values recorded for each fracture in a measurement zone. Consider gathering data on at least 50–100 fractures for each outcrop location or core. Consider these unstressed fractures or, in the case of filled fractures in core, a measure of width at the time of mineralization in the subsurface. This could be a poor assumption as fractures sometimes grow in width over time and mineral fill can develop a filling sequence or stratigraphy.
   b) Gather a measure of mechanical fracture aperture from BHI logs, most particularly FMI or CMI image logs. These are partially stressed apertures. The values are not particularly accurate but do show the relative order of magnitude of width and show relative differences with depth and by azimuth with respect to the *in-situ* stress state. These will generally be one to two orders of magnitude smaller than unstressed measurements.

c) Look for opportunities to gather fully stressed aperture measures from reservoir flow test data. In instances where you have open fracture spacing data from core or resistivity image observations and feel that over the period of the test all flow came from fractures rather than matrix, then apparent hydraulic fracture intensity can be calculated from the flow test permeability.

d) Consider running laboratory tests of fracture compressibility with stress on the rock units of interest. This will tie the approaches above together in a common permeability and aperture with stress curve. This can be done on both natural and simulated fractures (lapped saw cuts), see attached.

10) **Gather data on *in-situ* Stress Component magnitude and direction. Fractures parallel to Sh$_{max}$ are more permeable in the subsurface than fractures perpendicular to Sh$_{max}$ due to fracture aperture change or fracture compressibility.**

a) Create *in-situ* reservoir stress maps for the area of study or production

  i) Create maximum horizontal stress direction map (azimuth of Sh$_{max}$)

    1) Review the World Stress Map online and download Sh$_{max}$ azimuth for the area of interest or nearby areas; use this as a base for the independent stress mapping.

    2) Derive Sh$_{max}$ azimuth from Drilling Induced Fracture (DIF) patterns and Borehole Breakouts (BB) in BHI; Sh$_{max}$ must be parallel to DIF azimuth and perpendicular to BB azimuth.

    3) Derive Sh$_{max}$ azimuth from oriented or re-oriented core from DIF (Centerline Fractures and Petal Fractures).

  ii) Determine horizontal stress anisotropy difference (Sh$_{max}$-Sh$_{min}$) or stress ratio (Sh$_{max}$/Sh$_{min}$). This will dictate the relative importance of fracture azimuth on fracture permeability per set and/or fracture aperture by set for modeling.

    1) Determine Sh$_{min}$ values at wellbores

      a) Calculate Sh$_{min}$ from leak-off tests

      b) Estimate Sh$_{min}$ from static or dynamic Poisson's Ratio distribution and the integrated density log with depth (Sv)

11) **Fracture Scaling**

a) Natural tectonic fracture systems have long been known to be pervasive or spanning multiple orders of magnitude in development (9 OOM). These predominantly fault- and fold-related fractures have been shown to follow scaling law distributions. Particularly, Power Law and Fractal relationships have proven useful in defining scaling relations and in predicting scales of fractures not easily observed in our observational data sets. Properties such as fracture or feature length, spacing, and aperture or width have been shown to be fractal.

   i) Measure fracture spacing, size, and aperture (width) for small-scale fractures from core, BHI and, in some instances, thin section.

   ii) Measure fracture spacing, size, and aperture (width) for mid-scale fractures from core, and BHI logs. Fractures with larger aperture could be used as a guide to defining mid-scale fractures. Bed-contained fractures if they exist in core or BHI observations are a good fracture system to measure as a mid-scale fracture population.

   iii) Measure fracture spacing, size, and aperture (width) for large-scale fractures and/or fracture corridors from BHI logs and geophysical attributes. These will have large spacing and cut much of the stratigraphic section you observe or they occur as swarms of steeply dipping fractures 10s of feet wide or more, forming a system of great height.

   iv) For each measured parameter type, plot size vs frequency. Plot as a standard Frequency Diagram and look for a Power Law distribution. If yes, plot as a Fractal Diagram using frequency by size in log–log space. A linear trend over more than one log cycle indicates a fractal relation and the slope of linear fit is the Fractal Dimension.

   v) If not fractal, cross plot the various measured parameters and look for other forms of relation among the parameters that can be used to predict properties of unmeasured sizes of fractures.

   vi) Assign properties of length, height, width, and spacing to each fracture scale chosen to model.

12) **Predictions of properties away from control, including mechanical fracture predictions in 3D**

   a) In instances where there is little data from which to make direct observational measurements of fractures (little or no core, no appropriate outcrops, few if any BHI logs, or BHI of low image or interpretation quality) it may be necessary to predict fracture intensity throughout the reservoir.

     i) If enough numbers of sonic logs are available in the reservoir, predictions of relative fracture intensity can be made from rock mechanics moduli (G or M) in the manner described in previously of this Workflow.

     ii) Prediction of fracture intensity within a reservoir and across a structure can be made from strain maps. Assuming the rocks involved are brittle and that all strain goes into development of fractures, areas of high *in situ* strain should equate to areas of high-fracture intensity. Therefore, strain maps calibrated by some direct-fracture intensity measures can be used to quantitatively predict fracture intensity,

       1) Restoration Strain Maps – A common approach to structural mapping in the industry today involves validating structural cross sections and maps by kinematically restoring the data to a

pre-deformation geometry called a "restoration." This is done by hand or computer programs to sequentially unfault the faults and unfold the folds. If the restored section or map is not logical, changes to the interpretation are made and the restoration performed again, etc. As a further test, the restored section or map is forward modeled to recreate the original interpreted deformed geometry and adjustments made to create a result that looks like a final product.

In this process, the computer program can quantify the kinematic strain needed everywhere in the section or map to create that final product. These restoration strain sections and maps can be used as a guide to predict relative tectonic fracture intensity distribution in 2D or 3D. The assumption is that all strain goes into brittle fracture and no other deformation mechanisms were active when fracturing took place. The drawback to the approach is that it is a kinematic, and not dynamic, approach.

2) Numerical Modeling (Finite Element Modeling) – An important tool for mapping strain in the subsurface is either finite element or finite difference modeling. This is a dynamic and not kinematic approach to strain modeling. In this approach, a gridded model is created with physical mechanical properties assigned to every grid cell according to the original geometry of the undeformed state. Each rock layer is assigned its appropriate mechanical properties (measured or inferred) including a stress/strain curve with any indications of deformation mechanism changes during deformation. Pre-deformation ambient conditions are applied to simulate burial conditions at the time of deformation.

Displacements are made sequentially in the model to turn the undeformed grid block into the final observed geometry. Equivalent plastic strain, or another form of dynamic strain, is calculated for every grid cell for each displacement step. This calculation takes into account the forces surrounding each cell, and the mechanical behavior of each cell for each displacement step. A prediction can be made when each cell will fracture or if non-fracture ductile flow occurs. These dynamic strain maps can be excellent predictors of fracture intensity distribution and fracture properties for flow modeling. As with other approaches, there must be calibration with observed fracture distributions.

This approach is probably the best predictor of subsurface natural fracture prediction. However, it takes modeling expertise and lots of mechanical data and takes a lot of time and money to complete and validate.

13) **Calibration to production and testing**
   a) Map existing production and flow data
      i) Productivity Index (PI)
      ii) Initial Potential (IP)
      iii) Estimated Ultimate Recovery (EUR)
      iv) Kh
      v) Maximum Flow Rate
      vi) A Measure of Stimulation Success
      vii) Mud losses
      viii) Drilling rate and weight on string
      ix) In unconventional reservoirs wells showing unstimulated flow.
   b) Create high-value domains in the above data in map form and compare with fracture intensity maps.
   c) Cross plot values of the above at wells with fracture intensity for open + partially open fractures and calculate degree of correlation (r^2).

14) **Relation between fracture distribution and flow/permeability**
   a) Plot FIC for open + partially open fractures with Production Logs and look for correspondence to prove if fractures are controlling fluid flow and which fracture zones are actually contributing.
      i) Spinner Log
      ii) Temperature Log (especially for gas)
      iii) Assibilation Log (acoustic).
   b) Plot FIC for open + partially open fractures with the Stoneley Wave Log and look for areas where the Pin or Index Curve occurs in a high-intensity fracture zone.
   c) Correlate fracture zones from the FIC with Injector/Producer short-cuts or short-circuits, especially in horizontal wells.

15) **Create the 3D Visualization Diagram(s) for the Fractured Reservoir and Populate with the Various Scales of Fracture Features.**
   a) Draw perspective block diagrams for the reservoir that include size and spatial relationships and orientations for the various fracture features being modeled.
   b) Assign population statistics to size, spacing, width, and aperture of the various scaled feature sets being modeled.
   c) Consider creating multiple visualizations for separate elements of the model, as well as a composite visualization.
   d) To support more detailed sector or individual well flow simulations, consider smaller visualizations that honor local rather than field-wide fracture and reservoir data.

16) **Integrate micro-seismic surveys taken during hydraulic fracture stimulation stages with FIC derived from core or BHI logs.**
   a) Determine which oriented natural fracture swarms (open or closed) correlate with micro-seismic distributions within or outside of the treatment zone.

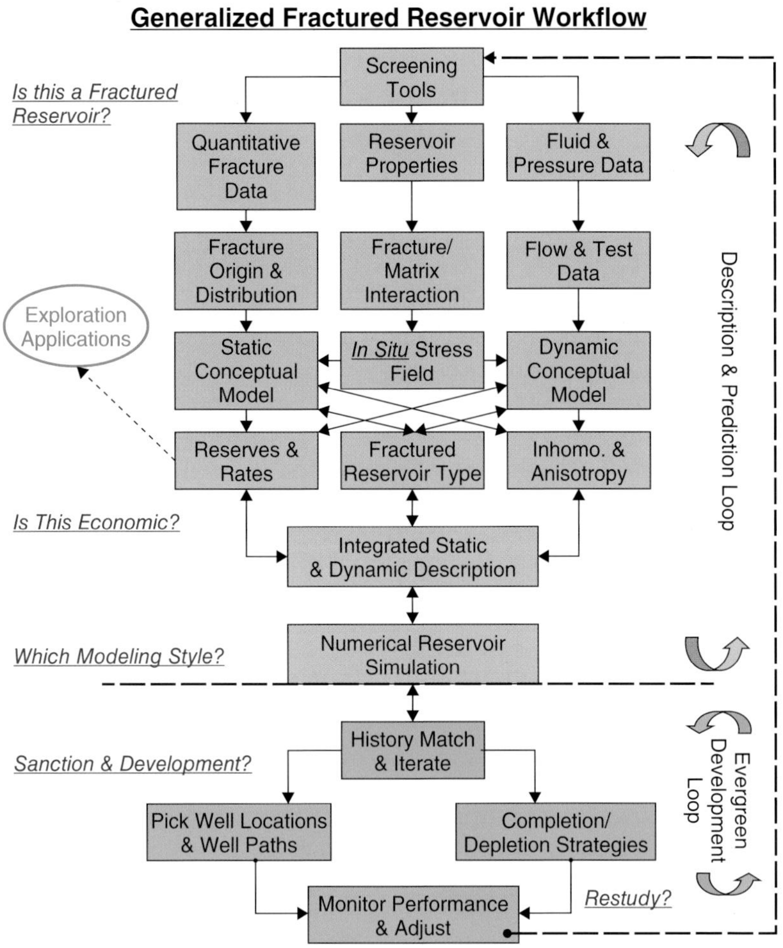

Generalized Fractured Reservoir Workflow

Figure A.1 Presented is a workflow for fractured reservoirs in general highlighting pre-sanction and post-sanction studies.

b) Look for stages where micro-seismic swarms indicate natural fracture re-activation outside of the target zone indicating non-containment.

17) **Determine Most Appropriate Fractured Reservoir Simulation Style**
   a) Type I – Single Porosity Simulation Model
   b) Type II – Dual Porosity Simulation Model
   c) Type III – Dual Porosity/Dual Permeability Simulation Model or Single Porosity Simulation Model with Permeability Anisotropy
   d) Type IV – Single Porosity Simulation Model with Strong Negative Permeability Anisotropy (Barriers to Flow)

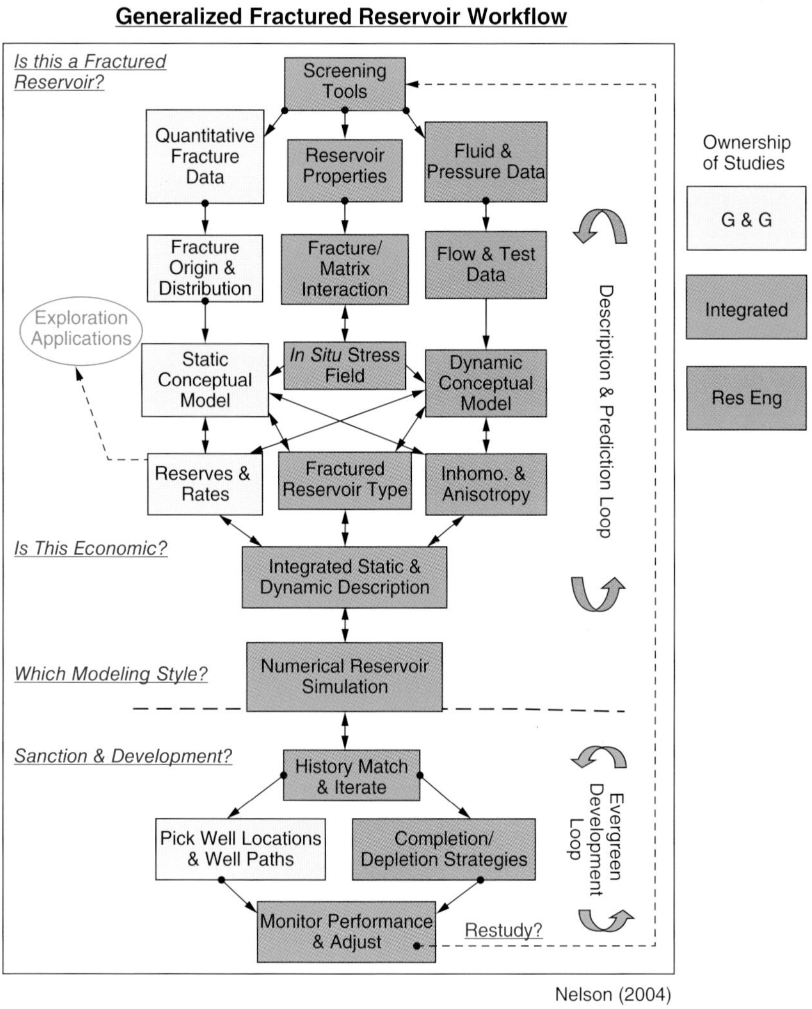

Generalized Fractured Reservoir Workflow

Nelson (2004)

Figure A.2 Presented is a workflow for fractured reservoirs in general highlighting the study areas owned by the geology and geophysics personnel, the engineering personnel, and those that are truly integrated studies.

18) **Create a Gridded Fracture Model for Simulation**

a) "Paint" grid cells directly in a 3D visualization to represent fracture flow properties as determined in the modeling.

b) Utilize an industry available or proprietary Fracture Modeling Software Package (FRACCA, FracMan, SVS, Petrel, etc.) to generate a gridded Continuous or Discrete, Deterministic or Stochastic Fracture Model, or a combination of the above.

Appendix B

How we Use Various Seismic Attributes to Predict Natural Fracture Intensity in the Subsurface, After Nelson (2006)

1) Fault and Structural Geometry from Structure Maps and Sections: Look for and define limits of:
   a) Significant outside and inside bends along normal fault traces.
   b) Fault displacement transfer zones.
   c) Fault tip-outs.
   d) Fault intersections.
   e) Hanging wall sides of normal and reverse faults.
   f) Up-dip corners of rotated fault blocks.
   g) Positive and negative curvature associated with folding.
   h) Zones or loci of high rate of change of dip.
   i) Zones within 100–200 m of major displacement faults.
2) Coherency Attribute Maps
   a) Low-coherency fracture zones related to faults.
   b) Low-coherency fault extensions where displacement becomes sub-seismic.
3) Dip Maps
   a) Areas of high dip rate associated with deformed zones.
   b) Abrupt changes in dip rate from high to low dip.
4) Dip Azimuth Attribute Maps
   a) Zones or lines of abrupt dip azimuth change.
   b) Greater amounts of azimuth shift may indicate greater degrees of fracturing.
5) Seismic Frequency Maps
   a) Zones that truncate the high end of seismic frequency.
   b) Zones where averaging frequency over the thickness of the unit in map form show relatively low frequency.
6) Root Mean Square (RMS) and Sum Absolute and Volume Amplitude Maps
   a) Zones of relatively low seismic amplitude indicating dissipation of energy in returning reflected seismic waves.
   b) Various forms of reflected wave low-returning energy.

Static Conceptual Fracture Modeling: Preparing for Simulation and Development,
First Edition. R.A. Nelson.
© 2020 John Wiley & Sons Ltd. Published 2020 by John Wiley & Sons Ltd.

7) Interval Velocity Maps
   a) Restricted zones of low-mapped interval velocity.
   b) Various forms of restricted zones where internal velocity is reduced.
8) Curvature Maps
   a) Zones of high curvature corresponding with zones of high-bending moments and high structural strain.
   b) High relative curvature in structural hinges associated with folds or faults.
9) Azimuthal Anisotropy Maps
   a) Inferred direction of the seismic fast direction.
   b) Zones where there is a relatively large difference between the seismic slow and fast directions.
   c) Zones where the azimuth of the fast direction changes rapidly.
10) P-impedance and S-impedance Maps
   a) Zones of both P-impedance and S-impedance display high relative values.
   b) Zones where S-waves are retarded.
   c) Zones where S-impedance is retarded but P-impedance less so.
11) Azimuthal Amplitude Vs Offset (AVO)
   a) Zones of low amplitude vs offset azimuth.
12) Edge Detection Maps
   a) Seismic discontinuities notably fault and fracture swarms.
13) Spectral Decomposition
   a) Zones of seismic attenuation as function of frequency from the spectra of each pulse.
   b) Zones of high Attenuation Coefficient.
14) 3D Curvature on Coherency
   a) Zones of high curvature associated with Coherency discontinuities.
   b) Areas where both positive and negative curvature are parallel and associated with Coherency discontinuities.
15) Ant Tracking (Schlumberger)
   a) Areas where automated discontinuity detection highlight "fault-like features" in seismic data. Areas are refined by allowing the user to interact with the data set by discarding and/or validating auto-picked features.
   b) Areas where the results appear like coherency and curvature on coherency.
16) Neural Networks using multiple attributes
   a) Areas where Neural Networks "trained" to predict fractures in 3D from numerous individual seismic attributes and calibrated with measured fracture intensity values from core or BHI.
17) Micro/Passive Seismic
   a) Areas where surface and downhole geophones resolve micro-seismic events during injection for either flooding and hydraulic fracturing

that form clouds of events or event lineations not associated with the actual hydraulic fracture.

b) Areas where the cloud of "events" show wide distribution in map and cross section indicating the reactivation of highly oriented natural fractures. These points can often depict the orientation and distribution of the natural fracture system through their linear trends in the event cloud.

18) Seismic Discontinuity Mapping

a) Areas or alignments of disruption on smoothed seismic surfaces with in bust in an applied auto-correlation program.

Appendix C

How I Learned to Interpret Natural Fractures in Core

When I was working on my dissertation on natural fracture development, I was studying at the Center for Tectonophysics at Texas A&M University. At the time, the Center was well known for its' multi-component approach to structural geology research with application to the O&G Industry; including outcrop, thin section, theoretical and experimental approaches. My dissertation involved all those aspects of study on a single topic. There were no significant core studies *per se* in the Center at the time. In addition, like all experimental rock mechanics laboratories at the time, rock-type variation was minimal with only one or two standard rock materials generally tested as surrogate for all sandstones, carbonates, and igneous rocks (shales were rarely tested).

However, upon graduation, I went to work at the Amoco Research Center to work on structural geology and fractured reservoirs, where most basic fracture and rock property data came from an abundance of subsurface core material (Amoco was the leading core-based reservoir description company pulling some 60 000′ of core per year in the 1970s). This was new to me and the only core description I had done in school was with Dr. Berg at Texas A&M in a stratigraphy/sedimentology class. That work was done on mounted core slabs (not the larger butts). One of my first studies at Amoco was on the Silurian Tuscarora Quartzite from Pennsylvania. The major observational data set was whole core material. I had to figure out how to work with this material that was usually the realm of the stratigraphers, petrologists and engineers. Indeed, there were few publications on fracture description in core and certainly no organized workflows. There were some studies published on natural fractures in drill cuttings. Most of the rest were based on outcrop studies and shape studies from seismic data.

So, I had to create an innovative approach from which to base our fracture studies. Early on, my emphasis was to find the zones and units in the core that displayed the natural fractures for reservoir properties (fracture porosity and fracture permeability). I recorded these best zones along with an idea or average fracture intensity and/or spacing and a rough idea of the percentage of

Static Conceptual Fracture Modeling: Preparing for Simulation and Development,
First Edition. R.A. Nelson.
© 2020 John Wiley & Sons Ltd. Published 2020 by John Wiley & Sons Ltd.

mineralized vs open fractures. I then shared these observations with the research petrologist working the core. He or she would usually find a petrologic reason why those fractures were well developed in that zone, often mineral composition, grain size, matrix porosity or diagenesis. This correlation gave a potential predictor of fracture distribution in later wells. Of course, this was shared with the research engineer working with the core generating petrophysical data for describing the fluid-flow capability of the reservoir. So, this became a truly multidisciplinary effort almost from day one.

Along the way, I worked with numerous Petrophysical Trainees with their year-long projects utilizing core. I showed them how to interpret whether natural fractures were important to their project or not, and show them how to further quantify them from a reservoir perspective.

After about five years of core descriptions, it became obvious that we needed an approach that was less qualitative and more quantitative to better develop a fracture model. I designed my own way of quantifying my observations. What is needed for fracture porosity and permeability equations is fracture spacing, or the average distance between parallel fractures. However, in vertical cores (we were not doing horizontal wells yet) with steeply dipping fractures a 4″ diameter core does not sample enough vertical fractures to get an accurate or even semi-constrained fracture spacing. Therefore, I decided to break the cored interval into regular depth intervals (1′, 1 m, ½ meter, etc.) and count and record the number of natural fractures in that sample interval. This approach is called fracture intensity and is inherently the inverse of fracture spacing. For modeling purposes, it is called a P_{10} measure. In addition, relative fracture orientation(s), fracture plane morphology and mineral fill type, and any running comments on fracture system origin were also recorded for that interval. From this data, I calculated a running fracture intensity curve that often-contained thousands of data points. This population of fracture intensities could then be treated statistically in terms of distribution statistics.

As I got better with this process and gained more experience (estimated well over 150 000′ of core interpretations) I found that others like John Lorenz and Scott Cooper were using a different recording technique by recording the along core distance between subsequent fracture occurrences. This gave another form of a fracture spacing measure. Every bit as good, but I prefer a regular sampling interval.

I use this quantification approach for both core and borehole image logs (BHI) interpretations. This allows for good comparison between the two, as well as comparison to standard wireline logs and production logging tool (PLT) response.

What I think is important in interpreting and quantifying natural fractures is to work with other professionals from different backgrounds and to be innovative in your approach.

References

Ameen, M.S., Buhidma, I.M., and Rahim, Z. (2010). The function of fractures and in-situ stresses in the Khuff reservoir performance, onshore fields, Saudi Arabia. *American Association of Petroleum Geologists Bulletin* 94 (1): 27–60.

Anderson, E.M. (1905). The dynamics of faulting. *Edinburgh Geological Society Transactions* 8: 393–402.

Anna, L.O., Pollastro, R., and Gaswirth, S.B. (2010). Williston Basin Province–Stratigraphic and Structural Framework to a Geological Assessment of Undiscovered Oil and Gas Resources. In: *U.S. Geological Survey Digital Data Series 69–W* (eds. L.O. Anna, R. Pollastro and S.B. Gaswirth), 17.

Barton, N. (2007). *Rock Quality, Seismic Velocity, Attenuation, and Anisotropy*. London: Taylor and Francis Group.

Bisdom, K., Gauthier, B.D.M., Bertotti, G., and Hardebol, N.J. (2014). Calibrating discrete fracture-network models with a carbonate three-dimensional outcrop fracture network: implications for naturally fractured reservoir modeling. *American Association of Petroleum Geologists Bulletin* 98 (7): 1351–1376.

Bosworth, W., Khalil, M., Clare, A. et al. (2014). Integration of outcrop and subsurface data during the development of fractured eocene carbonate reservoir at the East Ras Budran concession, Gulf of Suez, Egypt. In: *Advances in the Study of Fractured Reservoirs* (eds. G.H. Spence, J. Redfern, R. Aguilera, et al.), 333–359. London: Geological Society.

Bourbiaux, B., Basquet, R., Cacas, M.C., and Daniel, J.M. (2003). An Integrated Workflow to Account for Multiscale Fractures in Reservoir Simulation Models: Implementation and Benefits. Proceedings ATC 2003 Conference & Oil Show, Islamabad, Pakistan (3–5 October 2003).

Bourbiaux, B., Basquet, R., Cacas, M.C. et al. (2002). *An Integrated Workflow to Account for Multi-scale Fractures in Reservoir Simulation Models: Implementation and Benefits*. Society of Petroleum Engineers https://doi.org/10.2118/78489-MS.

Bratton, T., Canh, D.V., Cuc, N.V. et al. (2006). Nature of naturally fractured reservoirs. *Schlumberger Oilfield Review* 18 (2): 4–23.

Buckner, S., Nelson, R., Bayer, S. et al. (2013). Predicting Natural Fractures in Unconventional Reservoirs: Examples of Data Validation Techniques from the Bakken System, Mountrail County, North Dakota. Abstract of Talk, American Association of Petroleum Geologists Annual Convention, Pittsburgh, USA (19–22 May 2013).

Chopra, S. (2009). Interpreting fractures through 3D seismic discontinuity attributes and their visualization. *Canadian Society of Exploration Geophysicists Recorder* 34 (8).

Chopra, S. and Marfurt, K.J. (2007a). Volumetric curvature attributes add value to 3D seismic interpretation. *Leading Edge* 2007: 856–867.

Chopra, S. and Marfurt, K.J. (2007b). Seismic curvature attributes for mapping faults/fractures, and other stratigraphic features. *Canadian Society of Exploration Geophysicists Recorder* 32 (9).

Chorney, D., Smith, M., and Maxwell, S. (2016). Micro-seismic geomechanical interpretation of a Montney hydraulic fracture. *Canadian Society of Exploration Geologists Recorder* 41 (3).

Cioffi-Revilla, C. and Midlarsky, M.I. (2013). Power Laws, Spacing, and Fractals in the Most Lethal International and Civil Wars. Elsevier. https://ssrn.com/abstract=2291166 (accessed 19 June 2019).

Davy, P., Darcel, Le Goc, Munier, R. et al. (2018). DFN, why, how and what for, concepts, theories and issues. Second International Discrete Fracture Network Engineering Conference, Seattle, USA (20–22 June 2018). American Rock Mechanics Association Paper DFNE.

Fox, A., LaPoine, P., and Enachescu, C. (2009). Propagation of Uncertainty from Discrete Fracture Networks (DFNs) to Downstream Models. Presentation at the SPE ATW Upscaling of Fractured Carbonate Reservoirs in Hammamet, Tunisia (29 June–1 July 2009).

Gabrielsen, R.H., Nystuen, J.P., and Olesen, O. (2018). Fault distribution in the Precambrian basement in South Norway. *Journal of Structural Geology* 108: 269–289.

Gale, J.F.W. (2014). Natural Fracture Patterns and Attributes Across a Range of Scales. *American Association of Petroleum Geologists*. Search and Discovery, Article #41486, 1–40.

Geiger, S. and Matthai, S. (2014). What can we learn from high resolution numerical simulations of single-and multi-phase fluid flow in fractured outcrop analogues. In: *Advances in the Study of Fractured Reservoirs* (eds. G.H. Spence, J. Redfern, R. Aguilera, et al.), 125–144. London: Geological Society.

Gerhart, L.C., Anderson, S.B., and Fischer, D.W. (1990). Petroleum geology of the Williston Basin. In: *Interior Cratonic Basins*, vol. 51 (eds. M.W. Leighton, D.R. Kolata, D.F. Oltz and J. Eidel), 507–559. AAPG Memoir.

Gross, M.R. and Eyal, Y. (2007). Throughgoing fractures in layered carbonate rocks. *Geological Society of America Bulletin* 119 (11/12): 1387–1404.

Gross, M.R., Fisher, M.P., Engelder, T.L., and Greenfield, R.J. (1995). Factors controlling joint spacing in interbedded sedimentary rocks: integrating numerical models with field observations in the Monterrey formation, USA. In: *Fractography: Fracture Topography as a Tool in Fracture Mechanics and Stress Analysis* (ed. M.S.W. Ameen), 215–233. London: Geological Society.

Gross, M., Lukas, T.C., and Schwans, P. (2009). Application of Mechanical Stratigraphy to the Development of a Fracture-Enhanced Reservoir Model, Polvo Field, Campos Basin, Brazil. *American Association of Petroleum Geologists*. Search and Discovery, Article #20080, 1–2.

Gutierrez, M. and Youn, D.J. (2015). Effects of fracture distribution and length scale on the equivalent continuum elastic compliance of fractured rock masses. *Journal of Rock Mechanics and Geotechnical Engineering* 7 (6): 626–637.

Hanks, C.L., Lorenz, J., Lawrence Teufel, L., and Krumhardt, A.P. (1995). Lithologic and Structural Controls on Natural Fracture Distribution and Behavior Within the Lisburne Group, Northeastern Brooks Range and North Slope Subsurface, Alaska. *American Association of Petroleum Geologists Bulletin* 81 (10): 1700–1720.

Hart, B.S., Pearson, R., and Rawling, G.C. (2002). 3-D seismic horizon-based approaches to fracture-swarm sweet spot definition in tight-gas reservoirs. *The Leading Edge* 21 (1): 28–35.

Hencher, S.R. (2014). Characterizing discontinuities in naturally fractured outcrop analogues and rock core: the need to consider natural fracture development over geologic time. In: *Advances in the Study of Fractured Reservoirs* (eds. G.H. Spence, J. Redfern, R. Aguilera, et al.), 113–123. London: Geological Society.

Hennings, P.H., Olson, J.E., and Thompson, L.B. (2000). Combining outcrop data and three-dimensional structural models to characterize fractured reservoirs: an example from Wyoming. *American Association of Petroleum Geologists Bulletin* 84 (6): –849.

Hooker, J.N., Laubach, S.E., and Marrett, R. (2014). Universal power-law scaling exponent for fracture apertures in sandstone. *Geological Society of America Bulletin* 126 (9 & 10): 1340–1362.

Hubbert, M.K. and Willis, D.G. (1955). Important Fractured Reservoirs in the United States. 4th World Petroleum Conference Proceedings. Rome, Italy (6–15 June 1955).

Jamison, W.R. (2016). Fracture system evolution within the Cardium sandstone, Central Alberta foothills folds. *American Association of Petroleum Geologists Bulletin* 100 (7): 1099–1134.

Joubert, T.G. (1998). Optimal Drilling Direction in Folded Fractured Triassic Carbonates in Northeastern British Columbia Determined by Applying Fracture "Occurrence" to Frequency Intercept and Flow Diagrams. Master thesis. University of Calgary.

Khan, K., Al-Awadi, M., Dashi, Q. et al. (2009). Understanding Overpressure Trends Helps Optimize Well Planning and Field Development in a Tectonically Active Area in Kuwait. Presented at the SPE 2009 Asia Pacific Oil & Gas Conference & Exhibition in Jakarta, Indonesia (4–6 August 2009).

Kratz, M., Aulia, A., and Hill, A. (2012). Identifying Fault Activation in Shale Reservoirs Using Microseismic Monitoring during Hydraulic Stimulation: Source Mechanisms, b Values, and Energy Release Rates. *CSPG Recorder* 37 (6): 20–28.

Kulander, B.R., Dean, S.L., and Ward, B.J. (1990). *Fractured Core Analysis: Interpretation Logging, and Use of Natural and Induced Fractures in Core*, Methods in Exploration vol. 8. American Association of Petroleum Geologists.

Lacazette, A. (1991). New stereographic technique for the reduction of Scanline survey data of geologic features. *Computers and Geosciences* 17: 445–463.

Lacazette, A., Vermilye, J., Fereja, S., and Sicking, C. (2013). Ambient Fracture Imaging: A New Passive Seismic Method. Presented at the Unconventional Resources Technology Conference in Colorado, USA (12–14 August 2013).

Lange, A.G., Fourno, A., Delorme, M. et al. (2009). Workflow and Analysis Tools for the Characterization of Fractured Reservoirs. *American Association of Petroleum Geologists*. Search and Discovery, Article, #40424, 1–39.

Lorenz, J. C. (1995). Recognition and Use of Induced Fractures, and Other Features in Core Produced by the Coring Process. https://digital.library.unt.edu/ark:/67531/metadc794009/ (accessed 19 June 2019).

Lorenz, J. (2008). Fractured Reservoirs: Observation to Evaluation and Simulation. Education course. American Association of Petroleum Geologists. (15–17 September 2008).

Lorenz, J.C. and Cooper, S.C. (2018). *Atlas of Natural and Induced Fractures in Core*. Wiley.

Lorenz, J.C., Finley, S.J., and Warpinski, N.R. (1990). Significance of coring induced fractures in Mesaverde core, Northwestern Colorado. *American Association of Petroleum Geolists Bulletin* 74 (7): 1017–1029.

Mah, J., Samson, C., and McKinnon, S.D. (2011). 3-D laser imaging for joint orientation analysis. *International Journal of Rock Mechanics Mining Sciences* 48: 932–941.

Marrett, R. (1997). Permeability, porosity, and shear-wave anisotropy from scaling of open fracture populations. In: *Fractured Reservoirs: Characterization and Modeling* (eds. T.E. Hoak, A. Klawitter and P.K. Blomquist), 217–226. Rocky Mountain Association of Geologists.

Marrett, R., Ortega, O.J., and Kelsey, C.M. (1999). Extent of power-law scaling for natural fractures in rock. *Geology* 27: 799–802.

Massiot, C., McNamara, D.D., Nicol, A., and Townend, J. (2015). Fracture Width and Spacing Distributions from Borehole Televiewer Logs and Cores in the Rotokawa Geothermal Field, New Zealand. Presented at the World Geothermal Congress in Melbourne, Australia (19–25 April 2015).

McGinnis, R.N., Ferril, D.A., Morris, A.P., and Smart, K.J. (2017). Mechanical structural controls on natural fracture spacing and penetration. *International Journal of Rock Mechanics and Mining Sciences* 95L: 160–170.

McGinnis, R.N., Ferrill, D.A., Smart, K.J. et al. (2015). Pitfalls of using entrenched fracture relationships: fractures in bedded carbonates of the Hidden Valley fault zone, Canyon Lake Gorge, Comal County, Texas. *American Association of Petroleum Geologists Bulletin* 99 (12): 2221–2245.

Miller, R. (2014). Shale Reservoir Evaluation, Course notes. Presented at the American Association of Petroleum Geologists Annual Leadership Meeting in Houston, USA (5 April 2014).

Mynatt, I., Bergbauer, S., and Pollard, D.D. (2007). Using differential geometry to describe 3-D folds. *Journal of Structural Geology* 29: 1256–1266.

Nelson, R.A. 1975. Fracture permeability in porous reservoirs: an experimental and field approach. PhD dissertation. Texas A&M University, College Station, Texas, 1975, 171 p.

Nelson, R.A. (1983). Geologic Implications of Lineaments, Glen Canyon Area, Utah and Arizona - Structural Analysis. Presented at the 4th International Conference on the New Basement Tectonics in Oslo, Norway (10–14 August 1981).

Nelson, R.A. (1985). *Geologic Analysis of Naturally Fractured Reservoirs: Contributions to Petroleum Geology and Engineering No. 1*. Houston, Texas: Gulf Publishing Co.

Nelson, R.A. (2001). *Geologic Analysis of Naturally Fractured Reservoirs*, 2e. Houston: Gulf Professional Publishing/ Butterworth-Heinemann.

Nelson, R.A. (2002). Creating Static Conceptual Models for Fractured Reservoirs. American Association of Petroleum Geologists Proceedings. 2002 Annual Meeting in Houston, USA (10–13 March 2002).

Nelson, R.A. (2004). Integration of Fracture, Production, and Diagenetic Data in Static Subsurface Models Using 3D Visualization. Presented at the American Association of Petroleum Geologists Hedberg Conference, Structural Diagenesis: Fundamental Advances and New Applications form a Holistic View of Mechanical and Chemical Processes in Austin, USA (8–11 February 2004).

Nelson, R.A. (2006). Integrated Exploration & Evaluation of Fractured Reservoirs. Presented at the American Association of Petroelum Geologists Course in Houston, USA (8 –10 February 2006).

Nelson, R.A. (2010a). Characterization of Fractured Reservoirs: It's more than Just Fractures. Keynote Presentation, American Association of Petroleum Geologists. Presented at the Geoscience Technology Workshop, Role of Fracture and Geomechancial Characterization in the Hydrocarbon Industry: Middle Eastern Perspective in Rome, Italy (28–30 June 2010).

Nelson, R.A. (2010b). Key Elements to a Quality Fracture Model for Simulation: The Need for Multi-Scale Fracture Measurements to Constrain the Model. Presented at the European Association of Geoscientists and Engineers

Workshop on Naturally & Hydraulically Induced Fractured Reservoirs: From NanoDarcies to Darcies in Nafplio, Greece (10–13 April 2011).

Nelson, R.A. (2010c). Varying Role of Natural Fractures in Unconventional Reservoirs. Presented at the American Association of Petroleum Geologists Fall Education Conference in Houston, USA (7–8 October 2010).

Nelson, R.A. (2010d). Static Fracture Modeling: Key Requirements for a Quality Simulation Model. Presented at the Schlumberger World-Wide Web presentation in Kuwait (2 November 2010).

Nelson, R.A. (2011a). Evaluation and Quantitative Modeling of Fractured Reservoirs. Presented at the American Association of Petroleum Geologists Winter Education Conference in Houston, USA (13–14 February 2011).

Nelson, R.A. (2011b). Comparison of Data Sources to Quantitatively Constrain Fracture Intensity in Static Fracture Models; Translated. Presented at the Gulf Coast Association of Geological Societies Annual Meeting in Vera Cruz, Mexico (16 –19 October 2011).

Nelson, R.A., Lenox, L.C., and Ward, B.J. (1987). Oriented core: It's use, error, and uncertainty. *American Association of Petroleum Geologists Bulletin* 71 (4): 357–367.

Ortega, O., Marrett, R., and Laubach, S.E. (2006). Scale-independent approach to fracture intensity and average spacing measurement. *American Association of Petroleum Geologists Bulletin* 90 (2): 193–208.

Papadaki, E.S., Mertikas, S.P., and Sarris, A. (2011). Identification of linements with possible structural origin using Aster images and DEM derived products in Western Crete, Greece. *EARSeL eProceedings* 10: 9–26.

Paul, P., Zoback, M., and Hennings, P. (2007). Fluid Flow in a Fractured Reservoir Using a Geomechanically-Constrained Fault Zone Damage Model for Reservoir Simulation. Presented at the Society of Petroleum Engineers Annual Technical Conference and Exhibition in California, USA (11–14 November 2007).

Poppelreiter, M., Balzarini, M.A., De Sousa, P. et al. (2005). Structural Control on sweet-spot distribution in a carbonate reservoir: concepts and 3-D models (Cogollo Group, lower cretaceous, Venezuela). *American Association of Petroleum Geologists Bulletin* 89 (12): 1651–1676.

Prost, G.L. (1994). *Remote Sensing for Geologists, a Guide to Image Interpretation.* Amsterdam: Gordon and Breach Science Publishers S.A.

Qinghaul, E.I., Latham, J.P., and Tsang, C.F. (2017). The use of discrete fracture networks for modelling coupled geomechanical and hydrological behavior of fractured rocks. *Computers and Geotechnics* 85: 151–176.

Rawnsley, K. (2007). Structural and stratigraphic controls on fold-related fracturing in the Zagros mountains, Iran: Implications for reservoir development. In: *Fractured Reservoirs*, Special Publication. 270 (eds. L. Lonergan, R.J.H. Jolly, K. Rawnsley and D.J. Sanderson), 1–21. Geological Society London.

Richard, P., Lamine, S., Pattnaik, C. et al. (2017). Integrated Fracture Characterization and Modeling in North Kuwait Carbonate Reservoirs.

Presented at the Society of Petroleum Engineers International Petroleum Exhibition and Conference in Abu Dhabi, UAE (13–16 November 2017).

Ruehlick, B. (2015). From Borehole Images to Fracture Permeability and Fracturing Pressure. American Association of Petroleum Geologists. Search and Discovery, Article #41537, 1–33.

Segall, P. and Pollard, D.D. (1983). Joint formation in granitic rock of the Sierra Nevada. *Geological Society of America Bulletin* 94: 563–575.

Sonnenberg, S. and Weimer, P. (1993). Oil production from Niobrara formation, Silo field, Wyoming-fracturing associated with a possible wrench fault system (?). *Mountain Geologist* 30 (2): 39–53.

Stearns, D.W. (1968). Certain aspects of Fracture in Naturally Deformed Rocks. In: *NSF Advanced Science Seminar in Rock Mechanics* (ed. R.E. Ricker), 97–118. Bedford, Mass: Cambridge Research Laboratories.

Stiteler, T.C., Nelson, R.A., and Chacartegui, F.J. (1994). Fractured Reservoir Analysis with Examples from the Cretaceous Reservoirs of Lake Maracaibo, Venezuela. Presented at the 64th Society of Exploration Geophysicists Annual Meeting in Los Angeles, USA (27 October 1994).

Terzaghi, R.D. (1965). Sources of error in joint surveys. *Géotechnique* 13: 287–304.

Trice, R. (2000). Presentation at Naturally Fractured Reservoir Research Forum. Society of Petroleum Engineers Research Forum. Nice, France.

Wagner, P.D., Nelson, R.A., Lonnee, J.S. et al. (2010). Fracture Characterization of a Giant Unconventional Carbonate Reservoir, Alberta, Canada. Presented at the American Association of Petroleum Geologists International Conference and Exhibition in Calgary (12–15 September 2010).

Wagner, P.D., Nelson, R.A., Lonnee. J.S. et al. (2011). Natural Fracture Characterization of a Giant Unconventional Carbonate Reservoir, Grosmont Venture, Alberta, Canada: Implications for Recovery. Presented at the 2011 CSPG CSEG CWLS Convention in Calgary, Canada (9–12 May 2011).

Ward, M.V., Pearse, C., Jehanno, Y. et al. (2014). Machar oilfield UK Central North Sea: impact of seismic reprocessing of a complex fractured chalk field. In: *Advances in the Study of Fractured Reservoirs* (eds. G.H. Spence, J. Redfern, R. Aguilera, et al.), 361–377. London: Geological Society.

Waterhouse, M., Charara, M., and Nurmi, R. (1987). Focus on fractures. *Schlumberger Middle East Asia Reservoir Review* 3: 14–27.

Watkins, H., Healy, D., Bond, C.E., and Butler, R.W.H. (2018). Implications of heterogeneous fracture distribution and reservoir quality; an analogue from the Torridon Group Sandstone, Moine Thrust Belt, NW Scotland. *Journal of Structural Geology* 108: 180–197.

Weng, X., Kresse, O., Cohen, C.E. et al. (2011). Modeling of Hydraulic Fracture Network Propagation in a Naturally Fractured Formation. Presented at the Society of Petroleum Engineers Hydraulic Fracturing Technology Conference in Texas, USA (24–26 January 2011).

Zoback, M.D. (2007). *Reservoir Geomechanics*. New York: Cambridge University Press.

Index

Static Conceptual Fracture Modeling: Preparing for Simulation and Development,
First Edition. R.A. Nelson.
© 2020 John Wiley & Sons Ltd. Published 2020 by John Wiley & Sons Ltd.